Management and Industrial Engineering

Series editor

J. Paulo Davim, Aveiro, Portugal

More information about this series at http://www.springer.com/series/11690

Dhanasekharan Natarajan

ISO 9001 Quality Management Systems

Springer

Dhanasekharan Natarajan
Bangalore
India

ISSN 2365-0532   ISSN 2365-0540   (electronic)
Management and Industrial Engineering
ISBN 978-3-319-85381-9   ISBN 978-3-319-54383-3   (eBook)
DOI 10.1007/978-3-319-54383-3

Printed on acid-free paper

This Springer imprint is published by Springer Nature
The registered company is Springer International Publishing AG
The registered company address is: Gewerbestrasse 11, 6330 Cham, Switzerland

To Om Shakthi

Preface

ISO 9001 Standard specifies the requirements of a quality management system (QMS) for organization for delivering products and services with consistent quality to customers and satisfying applicable statutory and regulatory requirements. The QMS is applied by many organizations throughout the world.

Large organizations demand their suppliers to establish and maintain QMS with accreditation to ISO 9001. Government departments specify the accreditation requirement as precondition to participate in their tenders. Hence, it is necessary for all types of organizations to establish QMS conforming to ISO 9001.

Two categories of expertise are required for establishing a practical and effective ISO 9001 QMS. The functioning of organization and the design & manufacturing processes of products should be understood for preparing practical QMS documents. Secondly, the requirements of ISO 9001 should be understood for preparing the documents in accordance with the international standard. This book presents adequate guidance for understanding the requirements of the standard. The requirements are illustrated with examples from industries for unambiguous and clear understanding.

Chapter 1 introduces the internal customer–supplier relationship for understanding quality. The internal customer–supplier concept is regarded by some as one of the most powerful aspects to emerge from total quality management [1]. QMS is introduced using the concept for fulfilling the needs and expectations of internal and end customers with two examples.

The needs and the considerations for establishing the documents of QMS are indicated in Chap. 2. Process approach is one of the seven quality management principles of ISO 9001, and it integrates people with processes to achieve consistent results [2]. The approach is fundamental for preparing QMS documents. Process characterization is illustrated with examples to apply the process approach for preparing the documents. Sequence of processes, interactions of the processes and the categories of QMS documents are also explained.

Chapter 3 explains the management processes of ISO 9001. Guidance for identifying the internal and external issues that are relevant to organization and monitoring the issues is presented. Guidance is also provided for monitoring the needs and expectations of interested parties to provide products meeting customer requirements. The requirements for determining the scope and establishing the QMS of organization, leadership and planning are explained. Planning actions to address risks and opportunities and integrating the actions with QMS processes are illustrated for three operational processes.

Resources, monitoring and measuring resources, organizational knowledge, communication and documented information are grouped as support processes, and they are required for the operation of QMS processes. The requirements of the support processes are explained in Chap. 4.

Operational processes contribute significantly to the growth of organization. Planning, understanding customer requirements for products, design & development, control of external providers and production are the operational processes. The requirements of the operational processes and other related processes are explained with practical examples for deeper understanding in Chap. 5. Adequate information is provided for the processes related to the operational processes, and the processes are identification and traceability, customer property, preservation, post-delivery activities, control of changes, release of products and nonconformity control.

Chapter 6 describes the requirements of evaluating and improving the QMS of organization. Monitoring, measurement, analysis, evaluation, customer satisfaction internal audit and management review are explained, and examples are provided where required. Key performance indicators are suggested for measuring the performance of QMS processes.

The requirements of QMS should be amalgamated, i.e. integrated with the operational processes of the organization from the receipt of customer requirements to the delivery of products to customers for achieving and sustaining business growth. Integrating QMS requirements with ERP or other software is the most effective method, and it provides valuable benefits to organization. Software system engineering is briefly introduced. Requirements analysis with abbreviated activity diagrams and user acceptance testing is presented for providing inputs for developing software for the control of measuring resources in Chap. 7. The methods of integrating the QMS requirements with ERP software are illustrated for three operational processes in Chap. 8.

Thirukkural is an Indian classic literature in Tamil, one of the ancient languages of India. The classic literature is generally accepted as more than 2000 years old. The literature is secular, and some of the earliest translations in French (1848 and 1889), German (1856) and English (1886) indicate its universal acceptance. Thirukkural contains 1330 verses (kurals in Tamil). A kural is a couplet containing a complete and striking idea expressed in a refined and intricate metre [3]. Applying the ideas of the kurals for planning the actions to address risks and opportunities in QMS processes, design & development and production processes is presented in Chap. 9.

Bangalore, India   Dhanasekharan Natarajan

References

1. Slack N, Roden S (2015) Internal customer–supplier relationships. Wiley Encyclopedia of Management
2. ISO 9001:2015, How to use it, ISBN 978-92-67-10640-3, issued by iso.org
3. Rev. Dr. Pope GU (1886) The 'Sacred' Kurraḷ of Tiruvalluva-Nayanar, with introduction, Grammer, Translation & Notes. W.H.Allen & Co., London

Acknowledgements

Throughout my career, I was assisted by a team of committed and talented engineers and technicians with a lot of enthusiasm and initiatives. I would like to thank the team first for making me a better professional with time.

I would like to thank my wife, Rameswari, for encouraging me to engage in writing books after my retirement. I thank my son and daughter also for their encouragement.

Most importantly, I would like to express my sincere thanks to Venkataramani.R, Managing Director (Retd.) of Radiall Protectron, who provided competent persons for developing software system integrating QMS requirements with operational processes and for developing software for support processes. I would like to acknowledge the commitment and enthusiasm of the software team in developing the software. The efforts of the team were visible as error-free operations during the usage of the software system.

Contents

About the Author

Dhanasekharan Natarajan Electronics engineer from College of Engineering (presently Anna University), Guindy, Chennai, in 1970, obtained his postgraduate in Engineering Production (Q&R Option) from the University of Birmingham, UK, in 1984. He is a senior member (00064352-Member for 30 years), IEEE (MTT Soc.; Rel. Soc.; CS), and his biography is published by Marquis, USA, in their fourth edition, "Who's Who in Science & Engineering".

He retired as an assistant professor in RV College of Engineering, Bangalore. His earlier professional achievements at Bharat Electronics and Radiall Protectron include application of reliability techniques for defence equipments, root cause analysis on electronic component failures, qualification testing of electronic components as per US and Indian military standards, designing and implementing computerized quality management system, designing software for optical interferometer and design & manufacturing of lumped, semi-lumped and microwave cavity filters using self-developed software. He has authored two books and both are published by Springer.

The first book, A Practical Design of Lumped, Semi-lumped and Microwave Cavity Filters, Springer, 2013, provides the physical understanding of the terms of RF Filters and presents the design of lumped/semi-lumped and combline/iris-coupled cavity filters in tutorial form.

The second book, Reliable Design of Electronic Equipment: An Engineering Guide, Springer, 2015, presents the application of derating, FMEA, overstress analysis and reliability improvement tests for designing reliable electronic equipment and offers excellent support for electrical and electronics engineering students, bridging academic curriculum with industrial expectations.

Chapter 1
Introducing Quality Management System

Abstract Organization is a chain of internal suppliers and customers, who are collectively responsible to deliver finished products to end customer. The supplier-customer process relationship is used for understanding quality. The relationship is illustrated with two examples for understanding that quality is the fulfilment of the needs and expectations of internal and end customers. ISO 9001 quality management system (QMS) is introduced for fulfilling the needs and expectations of customers consistently.

1.1 Introduction

Organizations commit quality and business objectives for satisfying customers and other interested parties. The two objectives are not independent but complement each other for achieving them. Achieving both the objectives makes organizations grow spirally and the organizations become successful with sustained growth. A quality management system (QMS) should be established for achieving sustained growth. Understanding quality is essential for establishing the QMS of organization and it is derived from the process relationship between suppliers and customers.

1.2 Relating Internal and End Customers

Organization outputs products and end customers use the products. Products could be hardware or software or services or a combination of them. The term, product, implying hardware, software, and services, is used throughout the book.

Finished products relate organization and end customers. Organization implements a set of processes for delivering products to customers. Internal suppliers and customers exist within the organization and the internal processes relate them. The internal customer-supplier concept is regarded by some as one of the most powerful aspects to emerge from total quality management [1]. Internal customer-supplier

© Springer International Publishing AG 2017

D. Natarajan, *ISO 9001 Quality Management Systems*,

Management and Industrial Engineering, DOI 10.1007/978-3-319-54383-3_1

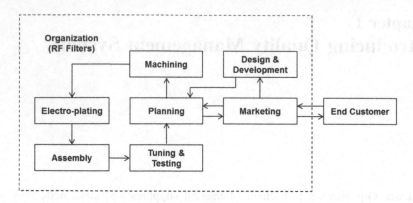

Fig. 1.1 Simplified process flow diagram for delivering RF Filters

relationship is explained for delivering RF Filters to end customers, using the simplified process flow diagram, shown in Fig. 1.1. The coordination of production processes by planning is not shown in the flow diagram to keep it simple.

Seven internal processes are shown in Fig. 1.1 for the organization involved in the design and manufacturing of RF Filters. The owner of a process is defined as internal supplier, who delivers semi-finished products to the owner of the next process. Those who receive inputs from internal suppliers are internal customers. For example, the owner of machining process (internal supplier) delivers mechanical parts to the next internal customer, Electro-plating. In general, process owner becomes internal customer for the previous process and the same owner becomes internal supplier to the next process. It could be visualized that organization is a chain of internal suppliers and customers, who are collectively responsible to deliver finished products satisfying the requirements of end customers.

1.3 Understanding Quality

The customer's perception of quality on products varies between customers and it could also vary for the same product of organization. Hence, it would be difficult to evolve a generic definition of quality. Understanding quality is more relevant than defining it. The philosophical description of quality [2] is appropriate for under-standing quality and some of the descriptions are:

- The pursuit of perfect that never ends,
- Quality is the fulfilment of needs,
- Quality is the degree of feeling happiness,
- Quality is the way of life.

The needs of customers include their expectations also. Fulfilling the needs and expectations of internal and end customers provide the feeling of happiness to the customers. Efforts to fulfill the needs and expectations should become a way of life in organization.

1.3.1 Needs and Expectations of Internal and End Customers

Two examples are provided for understanding that quality is the fulfilment of the needs and expectations of internal and end customers. The supplier-customer process relationship diagram shown in Fig. 1.1 for RF Filters is used for the illustration.

1.3.1.1 Needs and Expectations of End Customer

Marketing is one of the seven processes of organization, shown in Fig. 1.1. The process owner of Marketing is the internal supplier to end customer. The needs and expectations are identified using the process relationship between the internal supplier and end customer. They are documented for the end customer under product and service categories in Table 1.1.

1.3.1.2 Needs and Expectations of RF Filter Assembly

Figure 1.1 shows that the owner of Electro-plating process is the internal supplier to the next internal customer, Assembly. The process owner of Electro-plating process

Table 1.1 Needs and expectations of end customer

Category	End customer	
	Needs	Expectations
Product	Receiving RF filters as per customer orders	Honoring the commitments for delivery
		Consistent performance to eliminate incoming inspection
		Should withstand normal stresses during the assembly of RF Filters in equipment
		Stable and failure-free performance in equipment
Service	Appropriate actions in case of failures	Fast response for correcting failures
	Knowing the status of product delivery or service	Knowing the status of product delivery or service instantly

Table 1.2 Needs and expectations of RF filter assembly

Category	Internal customer: RF filter assembly	
	Needs	Expectations
Product	Receiving plated parts as per internal orders	Honoring the commitments for delivery
		Receiving plated parts with appropriate protection for preservation
		Consistent performance for defects-free RF Filter assembly and testing
Service	Correcting plating defects	Fast response for correcting plating defects

(internal supplier) is responsible to satisfy the needs and expectations of Assembly (internal customer). The needs and expectations are identified using the process relationship between Electro-plating and Assembly. They are documented for the internal customer, Assembly, under product and service categories in Table 1.2.

1.4   Quality Management System

Quality management system (QMS) is planned and established by documenting procedures for the processes of organization to fulfill the needs and expectations of internal and end customers. The international standard, ISO 9001, specifies the requirements of quality management system (QMS) to consistently provide products that meet customer and applicable statutory and regulatory requirements. The standard is applied by many organizations throughout the world.

ISO 9001 QMS processes are organized representing the four steps of Plan-Do-Check-Act (PDCA) cycle. Documented procedures and product related engineering documents are established for the processes integrating the needs and expectations of internal and end customers. General guidance for establishing the QMS of organization is provided in Chap. 2. QMS processes and the preparation of documented procedures for the processes are explained in subsequent chapters for organizations providing engineering products to customers. Appropriate changes could be made in the procedures for delivering software or services to customers.

References

1. Slack N, Roden S (2015) Internal customer–supplier relationships. Wiley Encyclopedia of Management
2. Borawski P (2011) Toward a definition of quality, American society for quality. Future of quality study

Chapter 2
Establishing ISO 9001 QMS Documentation

Abstract Every organization, however small it is, practices the requirements of a documented quality management system (QMS) for processing customer orders. The existing practices and the documents of the organization are improved and aligned to the requirements of ISO 9001 for establishing QMS. The benefits and the considerations for establishing the QMS are presented. Process characterization, sequence of processes and their interactions are fundamental for establishing QMS and they are illustrated with examples. Expertise required for preparing QMS documents and the needs for having quality manual for organization are also presented.

2.1 Documented QMS

Documentation refers to writing procedures or instructions for carrying out the processes of organization. Every organization, small or large, practices the requirements of a documented quality management system (QMS) for processing customer orders. Quotes for customer enquiries, order acceptance, availability of controlled engineering drawings, purchase orders on suppliers with appropriate purchasing information, identification system for the storage of products and displaying work instructions in manufacturing areas are some of the examples of practicing the requirements of documented QMS. The existing process procedures and the documents of organization are improved and aligned to the requirements of ISO 9001 for establishing QMS for the organization. The needs and the basic considerations for preparing QMS documents are explained.

2.1.1 Needs for Documentation

ISO 9001 QMS describes the minimum level of documentation for the QMS of organization. The needs for preparing the documentation are shown in Fig. 2.1 and they are:

© Springer International Publishing AG 2017
D. Natarajan, *ISO 9001 Quality Management Systems*,
Management and Industrial Engineering, DOI 10.1007/978-3-319-54383-3_2

5

Fig. 2.1 Needs for QMS documentation

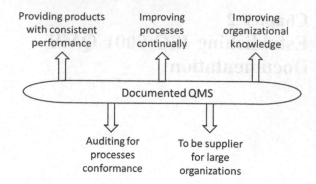

(i)   To provide products with consistent performance to customers.
(ii)   To improve processes continually.
(iii) To convert the expertise of individuals into organizational knowledge.
(iv)   To conduct audits of QMS processes for conformance.
(v)   To be a supplier for large organizations.

2.1.2 Basic Considerations

QMS documentation is prepared to comply with the requirements of ISO 9001 and to achieve the objectives of organization. Other considerations for preparing the documentation are shown in Fig. 2.2 and they are:

(i)   Optimum level of documentation:
      Optimum level of documentation is planned considering the size of orga-
      nization, the loss that would occur if processes were to fail and the com-
      petence of personnel.
(ii) Maintaining existing rhythm:
      Usually, QMS documentation is prepared for existing organization. The
      organization might have proven methods of managing processes to satisfy

Fig. 2.2 Considerations for preparing QMS documents

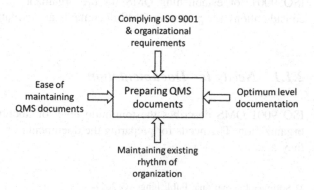

customers. The existing rhythm in managing processes should be maintained to the extent feasible in developing the QMS documentation as it provides comfort for understanding the documents and implementation. Innovative efforts are applied to avoid or minimize drastic changes in the existing methods as drastic changes might lead to quality problems resulting in customer dissatisfaction.

(iii) Ease of maintaining QMS documents:

QMS documents are amended as required by various corrective, preventive and improvement actions during the implementation of the QMS. Hence, they are organized for maintaining the documents easily. Needs to amend too many QMS documents for one action are avoided.

2.2   Process Approach

Process approach is one of the seven quality management principles of ISO 9001 and it integrates people with processes to achieve consistent results [1]. The approach is fundamental for preparing QMS documents. Process characterization is illustrated with examples to apply the process approach for preparing the documents.

2.2.1   Defining Process Characterization

Process characterization is defined as identifying elements and understanding their linkages for operations. The requirements relevant to processes are considered for characterizing the processes. The elements are listed below and the linkage between the elements is shown in Fig. 2.3:

(i)   Inputs
(ii)   Process procedure with acceptance criteria
(iii) Outputs
(iv)   Monitoring and measurement:

Monitoring and measurement activities are performed for controlling and improving processes. Additional information is provided for understanding monitoring and measurement.

2.2.2   Monitoring and Measurement

Monitoring is keeping track of a requirement and measurement is ascertaining (measuring) the level of achieving the requirement using gauges or test equipment or software. The requirement could be temperature, product delivery performance, time to carry out a process, etc. Generally, the requirements are monitored and

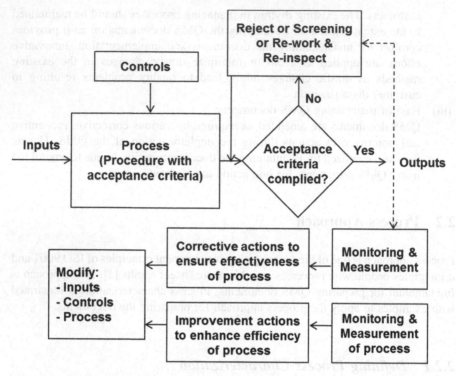

Fig. 2.3 Characterizing QMS process

measured simultaneously. Monitoring and measurement requirements are applied for control purpose, achieving intended process outputs (ensuring process effectiveness) and improving process (efficiency) [2].

2.2.2.1 Control Purposes

A sequence of operations is performed for carrying out a process. The relevant parameters of the operations are determined, monitored and measured for control purposes to minimize variations in the planned outputs of the process. The monitoring and measurement check points for control purposes are specific to process needs. Monitoring and measurement for control purposes are generally applicable for production processes.

2.2.2.2 Ensuring Process Effectiveness

A process is considered effective if the planned outputs of the process are achieved with consistent quality. Failures (defects) might be observed by process owners when carrying out processes or reported by other internal customers and end customers. The outputs of processes and the reported failures are monitored and measured using appropriate performance indicators. Corrective actions are implemented for the failures to ensure the effectiveness of processes. Monitoring and measurement for ensuring process effectiveness are explained in greater details in Sect. 6.2.

2.2.2.3 Improving Process Efficiency

Improving process efficiency is basically minimizing cost and time for carrying out processes. It is usually decided in management reviews. Process owners decide the methods of monitoring and measurement for improving the efficiency of processes.

2.3 Illustrating Process Characterization

Process characterization is well understood for product related production processes. Consider the reflow soldering process for mounting components on printed circuit boards (PCBs). The soldering process is characterized by inputs, process procedure with acceptance criteria and outputs. Monitoring and measurement are integrated with the process procedure. The elements of characterizing reflow soldering process are explained.

2.3.1 Inputs

Inputs for a process are the outputs of its previous process and the resources required for carrying out the process. Planning is the process previous to the reflow soldering process and it provides the inputs for reflow soldering. The inputs from planning are PCBs and surface mount components. The resources for carrying out the soldering process are reflow soldering equipment, tools, consumables, ultrasonic solvent bath for cleaning assembled PCBs and inspection equipment.

2.3.2 Process Procedure with Acceptance Criteria

Process procedure describes the steps for carrying out the reflow soldering process. The manufacturers of reflow soldering equipment describe the process procedure in detail. Baking PCBs, application of solder paste to pads, placing the parts and reflow soldering using the recommended profile are the major steps of the procedure [3]. The steps are described in the process procedure.

Acceptance criteria are specified in the process procedure to accept or rework or reject the assembled PCBs. Cleanliness and uniform solder joints are examples of acceptance criteria [3]. The PCBs are visually inspected after soldering and cleaning using the specified inspection equipment. In-process visual inspection for PCBs after applying solder paste to pads could also be specified in the process procedure. Quality needs of the internal customer of reflow soldering process are addressed in the process procedure.

2.3.3   Quality Needs of Internal Customer

The assembled PCBs after reflow soldering are delivered to the internal customer for electrical testing of the PCBs. Some of the critical quality needs of the internal customer are that CMOS components should be free from ESD related damages and the electrical performance of chip components should not have been degraded due to thermal shock in solvent cleaning process. The quality needs of the internal customer are addressed in the process procedure for reflow soldering process. Precautions to prevent ESD damages and for solvent cleaning of assembled PCBs are specified in the process procedure.

2.3.4   Monitoring and Measurement

The reflow soldering process parameters are monitored and measured to control the process for maximizing the output of acceptable assembled PCBs without rework i.e. for improving first time yield. Examples of process parameters are PCB baking temperature, soldering profile and the status of cleaning solvent. The requirements of monitoring and measuring process parameters for control purposes are addressed in process procedure.

The soldering defects observed during the inspection of assembled PCBs and those reported by the internal customers are monitored and measured. The defects are analyzed to identify the root causes of the defects and corrective actions are implemented to ensure process effectiveness. Improving process efficiency (minimizing cost and time for reflow soldering) through monitoring and measurement is a technical assignment and it is not explained.

2.3.5   Outputs

The outputs of reflow soldering process are assembled PCBs for the next internal customer and quality records. The outputs are:

(i) Acceptable assembled PCBs without visual defects and electrical damages to parts.
(ii) Quality records to provide evidences for carrying out the process:

  – In-process and final inspection records
  – Monitoring and measurement results as specified in the process procedure.

2.4 Process Approach for Controlling Engineering Documents

Process characterization is explained addressing the requirements of engineering documents. Creating, updating and controlling are the requirements the documents. Adequate information is provided in Sect. 4.8 for preparing the quality procedure addressing the requirements of engineering documents. The requirements that are unique for the quality procedure are presented.

2.4.1 Inputs and Outputs

The design and development outputs of product are the inputs for creating engineering documents. The resources are software drafting systems. The requirements of the quality procedure are implemented is applied during the preparation of the documents. The outputs of the quality procedure are:

  (i) Approved engineering documents
  (ii) Master list of engineering documents
  (iii) Distribution or access list for the documents
  (iv) Change control information for updated engineering documents
  (v) Retaining obsolete engineering documents with suitable identification as specified in process procedure.

2.4.2 Quality Needs of Customers and Acceptance Criteria

Process procedure should address the quality needs of internal and end customers. Although the design team of products controls engineering documents, they are not the users of the documents. End customers, production departments, external providers of products and quality assurance are the users of the documents. Avoiding contradicting engineering information in documents and providing information for achieving the planned results of production processes are examples of the quality needs of the customers. The quality needs of the customers should be addressed in the quality procedure, as appropriate.

Engineering documents are reviewed before approval. The review requirements are part of the quality procedure and they are listed in Sect. 4.8.1. The review requirements serve as acceptance criteria and controls for the process procedure.

2.4.3   Monitoring and Measurement

Errors in engineering documents might be reported by internal and end customers. Design team reviews the errors and updates relevant engineering documents. The number of updates indicates the effectiveness of preparing and reviewing engineering documents by design team. Monitoring and measurement for minimizing time (improving process efficiency) for preparing engineering documents are decided in management reviews.

2.5   QMS Processes

QMS processes are those specified in ISO 9001 and those determined by organization for their QMS. ISO 9001 processes are related to management, support, operation, performance evaluation and improvement. Examples of the additional processes determined by organization are:

(i)   Failure mode and effects analysis for product design
(ii)   Statistical techniques for manufacturing
(iii)   Reward schemes for improving processes continually
(iv)   Periodical reviews of customer orders to satisfy their requirements
(v)   Additional controls for contract manufacturing

QMS processes are characterized and documented (Sects. 2.3 and 2.4). The sequence of the processes and their interactions are also determined and documented.

2.5.1   Sequence and Interaction of QMS Processes

Sequence of processes is the order of implementing the processes for achieving planned results. Most of the processes receive inputs from many other processes and their outputs are used by multiple processes during implementation of the processes. In other words, there is interaction or linkage between the processes. Sequence and interaction of processes are illustrated with an example.

2.5.1.1 Illustration

A simplified process flow diagram showing the sequence and interactions of management, support and some of the product related processes of RF Filters are shown in Fig. 2.4. The sequence of production processes are in series. It can be seen in the figure that the process, Tuning and Testing for RF Filters, interacts with management, support and product related processes. Additional information for planning and establishing interaction between processes is presented for ensuring process effectiveness.

2.5.1.2 Additional Information for Process Interactions

Timing and sequence of interacting processes should be considered for determining the processes [4]. As an example, consider ultrasonic cleaning process that interacts with RF Filter Assembly process. The cleaning process is not shown in Fig. 2.4. The ultrasonic solvent cleaning bath is maintained at appropriate intervals for removing the flux residues of PCBs effectively. The timing of interaction for maintaining the bath is decided considering the volume of PCBs in the assembly process.

Planning process interacts with many other processes such as marketing, design, external providers, verification of externally provided products and production processes. The sequence of receiving inputs from the interacting processes are determined and documented for ensuring the effectiveness of planning process.

2.6 QMS Documents

QMS documentation is usually developed as multi-level documents to protect the confidentiality of engineering and other documents of organization. Documentation begins with establishing product related engineering documents and it is continued with the preparation of quality procedures for the processes of organization. The approach ensures clarity and continuity in the presentation of QMS documents. Both

Fig. 2.4 Simplified sequence and interaction of processes for RF filters

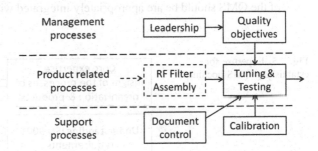

upward and downward linkages are ensured in the preparation of the documents. Although quality manual document is not mentioned in ISO 9001, it is necessary for an organization. The reasons for preparing the document are explained.

2.6.1 Quality Manual

Quality manual is a document that presents the overview of ISO 9001 QMS of organization. The quality manual is required for:

(i) Quality manual presents not only the overview of ISO 9001 QMS but also addresses business capabilities and briefly other abilities to achieve the planned results of QMS, thus supporting the business needs of organization.
(ii) Potential customers would like to understand the QMS of organization and plan assessment audit prior to visiting the organization. Quality manual supports the planning of assessment audit. Similar requirement might be expressed by certification bodies before registration audits.
(iii) All the clauses of ISO 9001 should be addressed in the QMS documents of organization. Documented information (quality procedures) is not needed for all the processes of ISO 9001. Quality manual is the means to ensure that all the clauses of ISO 9001 are addressed.

It is recommended that the quality manual is preferably prepared after the trial implementation of the QMS of organization and updating the documents based on the observed non-conformities. The approach ensures clarity in quality manual.

2.6.2 Expertise for Preparing QMS Documents

Two categories of expertise are required for preparing practical QMS documents. The requirements of expertise are shown in Fig. 2.5 and the requirements are:

(i) Core expertise:
ISO 9001 QMS does not specify stand-alone requirements. The requirements of the QMS should be are appropriately integrated with the operational process

Fig. 2.5 Expertise for establishing QMS documents

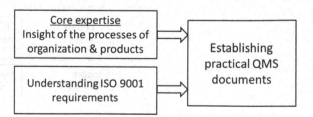

of organization. The insight into the functioning of organization and the design and manufacturing processes of products are the core expertise required for preparing practical QMS documents for organization.

(ii) Understanding ISO 9001:

Understanding the requirements of ISO 9001 is essential for preparing QMS documents in accordance with the international standard. The requirements of ISO 9001 are explained in the next four chapters. Adequate guidance and examples from industries are provided to facilitate the understanding the requirements.

References

1. ISO 9001 (2015) How to use it, ISBN 978-92-67-10640-3, issued by iso.org
2. The Process Approach in ISO 9001 (2015) ISO/TC 176/SC 2/N1289, issued by iso.org
3. Bergenthal J, Reflow soldering process considerations for surface mount applications, Kemet Electronics Corporation, F2102A, Oct. 97
4. ISO/TC 176/SC 2/N 544R3 (2008) ISO 9000 Introduction and support package: guidance on the concept and the use of the process approach for management systems

Chapter 3
Management Processes

Abstract The benefits of implementing QMS are presented. External and internal issues could affect the ability of organization to achieve the intended benefits of QMS. Guidance for identifying the internal and external issues that are relevant to organization and monitoring the issues is presented. Guidance is also provided for monitoring the needs and expectations of interested parties to provide products meeting customer requirements. The requirements for determining the scope and establishing the QMS of organization, leadership and planning are explained. Planning actions to address risks and opportunities and integrating the actions with QMS processes are illustrated for three operational processes.

3.1 Introduction

Implementing ISO 9001 quality management system (QMS) has become a necessity for organizations. The benefits of implementing the QMS for organization are [1]:

(i) Assessing the overall context of organization for stating objectives and identifying new business opportunities.

(ii) Consistently meeting and exceeding the expectations of customers, resulting in new clients and increased business.

(iii) Increasing productivity and efficiency, bringing internal costs down.

(iv) Meeting statutory and regulatory requirements.

(v) Expanding into new markets, as some clients require ISO 9001 before doing business.

(vi) Identifying and addressing risks associated with organization.

The commitment of top management is essential for establishing, implementing, maintaining and continually improving the effectiveness of ISO 9001 QMS. Top management demonstrates their commitment through leadership and planning for the QMS. Context of the organization, leadership and planning are grouped as management processes. Adequate information and examples are provided for understanding and documenting the processes.

© Springer International Publishing AG 2017   17
D. Natarajan, *ISO 9001 Quality Management Systems*,
Management and Industrial Engineering, DOI 10.1007/978-3-319-54383-3_3

3.2 Organization and Its Context

The requirements of ISO 9001 for understanding the context of organization are not new to the management of the organization. The external and internal issues that are relevant to the purpose (business objectives) and strategic direction of organization are identified and discussed in board meetings. The decisions are recorded for implementation. The status of implementing the decisions is reviewed in subsequent board meetings. Similar approach is applied for monitoring the external and internal issues by the process owners of organization except that the issues are assigned by top management to the process owners for monitoring.

3.2.1 Documentation Requirements

The intended benefits of ISO 9001 QMS (Sect. 3.1) might not be achieved even with the continued commitment of top management for the QMS. The ever changing external and internal issues might inhibit or deteriorate the achievement of intended benefits of ISO 9001 QMS. Hence, organization needs to identify relevant external and internal issues and monitor them to achieve the benefits. The suggested generic procedure for determining, monitoring and reviewing the issues are shown in Fig. 3.1.

3.2.1.1 The Procedure

The external and internal issues relevant to organization are determined during management reviews. Top management assigns monitoring the issues to appropriate process owners for monitoring. The process owners monitor the issues and prepare reports indicating the methods and the results of monitoring. If feasible, appropriate metrics are also decided for presenting the results. The reports are inputs for subsequent management reviews. The results of the analysis are evaluated in management reviews to decide the course of corrections. Quality manual is appropriate for documenting the procedure for understanding the organization and its context. The reports of process owners and the decisions of management reviews serve as documented information (quality records), providing evidences for monitoring and reviewing the relevant issues.

Fig. 3.1 Monitoring external and internal issues

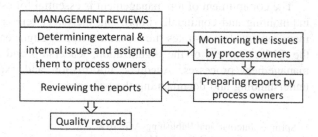

3.2.1.2 Suggested List of External and Internal Issues

ISO 9001 suggests a list of external and internal issues, including legal issues. Developments in statutory and regulatory requirements are external legal issues. Organization could identify relevant external and internal issues from the list for monitoring. Safety in operations is an internal issue and it could also be considered. Monitoring external and internal issues are explained with examples for better understanding.

3.2.2 Monitoring External and Internal Issues

Technology is an external issue for organization. Technology is changing with time due to inventions in materials and processes. The existing products of organization might become obsolete with changes in technology. Design team is assigned to monitor technology developments and submit half yearly reports to management. Collecting information on technology development from literature and on the products of competitors is part of monitoring. A report is prepared by the design team indicating the methods of monitoring and the results of analysis. The reports are inputs for management reviews.

3.2.2.1 Internal Issues

The performance of organization is an internal issue for organization. Any deterioration in the performance affects business growth and the satisfaction level of customers. Measuring growth of revenue and profit, level of customer satisfaction, business gain with new customers and business lost with existing customers are some of the parameters that could be monitored and measured for the internal issue. Monitoring and measuring the performance of organization are assigned to marketing team. Software support is essential for the monitoring and measurement. A report is prepared by the marketing team indicating the methods of monitoring and the results of analysis. The reports are inputs for management reviews.

3.3 Needs and Expectations of Interested Parties

Interested parties are those who affect or have the potential to affect the organization's ability to consistently provide products that meet customer and applicable statutory and regulatory requirements. The actions that are planned by organization are:

Fig. 3.2 Monitoring needs
and expectations of interested
parties

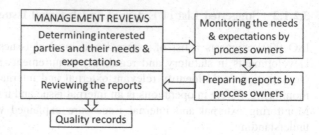

(i)   Determining the interested parties.
(ii)   Determining the needs and expectations of the interested parties.
(iii) Monitoring and reviewing the needs and expectations of the parties.

Examples of interested parties with their needs and expectations are presented. The suggested generic procedure for addressing the needs and expectations is shown in Fig. 3.2.

3.3.1 Interested Parties

Interested parties are shareholders, customers, employees, unions, suppliers and society [2]. The customer who contributes more than 50% of sales is an example of interested party for organization. Single source supplier providing critical parts is another example of interested party. Organization determines the interested parties that are relevant to their QMS.

The list of interested parties should also include the parties who have potential to affect the ability of organization for meeting their requirements. For example, if the effluents (air, water, etc.) of organization are not controlled, society, supported by regulatory agencies, has the potential to affect the ability of organization and hence, society, if relevant, should be included in the list of interested parties.

3.3.2 Requirements of Interested Parties

The needs and expectations are the requirements of interested parties. Needs are the necessities and the expectations are forecasts or predictions. For example, the needs of external provider (supplier) are the receipt of purchase orders from organization. The expectation of the supplier could be technical assistance from organization or higher volume of business in the future. The needs and expectations that are relevant to the requirements (business objectives) of the QMS of organization should be determined. For example, providing technical assistance to supplier should benefit organization and the supplier in improving the quality and delivery of externally provided products.

3.3.3   *Determining, Monitoring and Reviewing*

Management review is one of the forums to determine interested parties and the requirements of the parties. Synergizing the expertise of process owners is a practical method to identify the interested parties and their requirements. The identified requirements of interested parties are assigned to process owners for monitoring.

A formal report is prepared by the process owners indicating the method of monitoring the requirements of interested parties and the results of monitoring. If feasible, appropriate metrics are also decided for presenting the results. The results of the analysis are evaluated in management reviews to decide the actions.

Quality manual is appropriate for documenting the procedure for understanding the needs and expectations of interested parties. The reports of process owners and the decisions of management reviews serve as documented information (quality records), providing evidences for monitoring and reviewing the requirements of interested parties.

3.4   Scope of Quality Management System

The scope of organization is a set of statements defining the operational boundaries of the organization. It is established considering the present and long term requirements of organization. The considerations for establishing the scope of organization are shown in Fig. 3.3 and the considerations are:

(i)   Products offered:
       The broad categories of products offered by the organization are indicated.

Fig. 3.3 Inputs for defining
the scope of organization

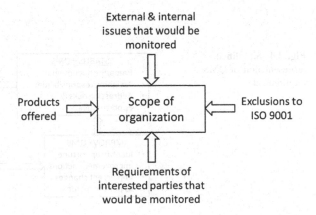

(ii)   External and internal issues that would be monitored (Sect. 3.2):
        The issues relevant to the QMS of organization are included in the scope
        appropriately.
(iii)   Requirements of interested parties that would be monitored (Sect. 3.3):
        The requirements of interested parties relevant to the QMS of organization
        are addressed in the scope appropriately.
(iv)   The applicability of all the requirements of ISO 9001:
        The requirements of ISO 9001 are examined whether all of them are
        applicable for organization or not. The exclusions to ISO 9001 are indicated
        with justifications.

Quality manual is appropriate for documenting the scope of QMS. The scope of
QMS is maintained i.e. reviewed and updated as needed by organization. The scope
of QMS states the types of products offered and the exclusions to ISO 9001 with
justifications. The QMS of organization is considered conforming to ISO 9001,
only if the exclusions do not affect the organization's ability or responsibility to
ensure the conformity of its products and the enhancement of customer satisfaction.

3.5   Quality Management System and Its Processes

The QMS of organization is a set of processes that are integrated to provide
products satisfying customer and applicable statutory and regulatory requirements.
The processes are those specified in ISO 9001 and those determined by the organization. Examples of processes determined by organization are provided in
Sect. 2.5. Organization establishes, implements, maintains and continually
improves its QMS. The requirements for establishing, implementing, maintaining
and improving the QMS of organization are explained. The simplified representation for the requirements of QMS is shown in Fig. 3.4.

Fig. 3.4 Simplified
representation of QMS
requirements

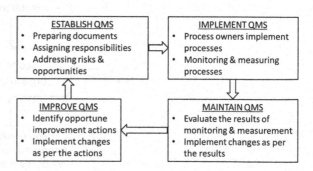

3.5.1 Establishing QMS

The QMS of organization is established by preparing the documents of the QMS and assigning responsibilities for implementation. The activities for establishing the QMS are:

  (i) Preparing quality manual.
 (ii) Preparing generic and product related procedures by characterizing QMS processes as indicated in Sect. 3.5.1.1.
(iii) Assigning responsibilities and authorities for implementing the processes.
 (iv) Addressing risks and opportunities in the processes of QMS.

3.5.1.1 Characterizing QMS Processes

Examples are provided in Chap. 2 for characterizing QMS processes. The requirements for characterization are:

  (i) The inputs required for the processes and the outputs from the processes (Sects. 2.3.1, 2.3.5 and 2.4.1).
 (ii) The sequences and the interaction of the processes (Sect. 2.5.1).
(iii) The methods of operation with acceptance criteria (Sects. 2.3.2 and 2.4.2).
 (iv) Monitoring and measurements for controlling and ensuring the effectiveness of the processes (Sects. 2.3.4 and 2.4.3).
  (v) The inputs for processes include the resources needed for the operation of the processes. The availability of the resources is ensured for carrying out the processes.

3.5.1.2 Assigning Responsibilities and Authorities

Roles, responsibilities and authorities are defined by top management. Section 3.6.4 could be referred for defining and documenting them. Process owners are responsible for implementing their processes as per the relevant QMS documents of processes.

3.5.1.3 Addressing Risks and Opportunities

Addressing risks and opportunities in QMS processes is explained in Sect. 3.7. Identifying the actions to address the risks and opportunities and integrating the actions with QMS processes are illustrated for three QMS processes in Sects. 3.8–3.10.

3.5.2   Implementing, Maintaining and Improving QMS

Process owners implement QMS processes as per established the documents of the processes. Monitoring and measurements are performed during the implementation of the processes as specified in the documents The QMS is maintained by:

 (i) Evaluating the processes based on the results of monitoring and measurement.
(ii) Implementing the changes in QMS documents as per the findings (corrective actions) of evaluation to achieve the intended results of processes.

When evaluating the processes, opportune improvement actions are also identified. QMS documents are amended to improve the processes. Performance evaluation and improvement of QMS processes are discussed in Chap. 6.

3.5.3   Maintaining and Retaining Documented Information

Documented information for the QMS of organization is maintained to support the operations of the QMS. Quality manual, quality procedures for the required generic QMS processes, product related engineering documents and the required external standards are maintained.

3.5.3.1   Retaining Documented Information

Quality records are referred as retaining documented information. The records provide evidences to demonstrate that the processes of organization are being carried out as planned in QMS documents. Quality records are one of the outputs of processes and the requirements of quality records are specified in the documents of the processes.

3.6   Leadership

Leadership is the influence exerted by one member to change, shape and direct the action of other members in organization [3]. Top management is entrusted to provide the leadership to organization for improving the business of organization and customer satisfaction. Top management plans appropriate actions and organizes for the implementation of the actions, demonstrating their accountability for achieving the business objectives of organization. Similar leadership efforts are required from top management for achieving the objectives of QMS.

3.6.1 Leadership and Commitment

ISO 9001 explicitly specifies a set of actions to demonstrate the leadership and commitment of top management for implementing, maintaining and continually improving the QMS of organization. Evidences for implementing the actions should be documented and the documents provide practical information for managing risks to the advantage of organization. The documentary evidences demonstrate the leadership of top management and they would be preferred by certification bodies.

3.6.1.1 Taking Accountability

Accountability for the effectiveness of QMS is similar to that of achieving the business objectives of organization. The accountability for QMS could be included in defining the responsibilities of top management.

3.6.1.2 Establishing Quality Policy and Quality Objectives

Top management establishes quality policy (Sect. 3.6.3) and quality objectives (Sect. 3.11.1) for the QMS considering the context (Sect. 3.2) and strategic direction of organization. They are documented and communicated within the organization.

3.6.1.3 Integrating QMS with Business Processes

Business processes are the set of activities defined in product related engineering documents for converting customer orders into deliverable products. The requirements of QMS are integrated appropriately in the quality procedures and product related engineering documents of organization for implementation. Integrating the requirements of QMS with business processes using ERP software would be more effective. Chapters 7 and 8 provide additional information for the integration.

3.6.1.4 Promoting Process Approach and Risk-Based Thinking

Process approach is ensured in the preparation of quality procedures and product related work instructions for production operations. Risk based thinking is aimed at identifying preventive actions for QMS processes. Top management reviews the effectiveness of actions taken to address risks and opportunities for QMS processes.

The decisions and actions are recorded. During management reviews, top management encourages process owners to apply risk based thinking in their processes.

3.6.1.5 Ensuring the Resources for QMS

The availability of resources for achieving the planned results of QMS is ensured. The additional resource needs for implementing QMS are identified during management reviews and they are documented for further actions. Documents are maintained indicating the capabilities of existing resources (people, infrastructure and environment) of organization.

3.6.1.6 Communicating QMS Requirements

The importance of effective quality management and of conforming to the QMS requirements is communicated within organization. Quality policy, quality objectives and the trends in the quality objectives could be displayed in the functional areas of organization.

3.6.1.7 Achieving the Intended Results of QMS

The QMS of organization is reviewed by top management at planned intervals to ensure that the QMS achieves its intended results. The records for the inputs, discussions and outputs of management reviews serve as evidences.

3.6.1.8 Engaging, Directing and Supporting the Effectiveness of QMS

Effective leaders engage in actions that inspire confidence in others and encourage these persons to accept their influence [3]. Top management should engage in actions such as application of process approach, risk based thinking and the actions that contribute for the effectiveness of QMS. Top management could record such actions in their reviews with management personnel and ensure that the actions are implemented. The actions of top management influence others to accept management directives and support for ensuring the effectiveness of QMS.

3.6.1.9 Promoting Improvement

The outputs of management review provide evidences for promoting the improvement of QMS. Reward schemes for improvements within organization could also be planned.

3.6.1.10 Supporting Other Management Roles

Apart from the above set of actions, top management could decide additional support roles for improving the QMS. The creation of cross functional team for measuring customer satisfaction through visits to customer organizations is an example for demonstrating leadership and commitment towards QMS.

3.6.2 Customer Focus

Top management demonstrates leadership and commitment towards customer focus within organization. Some of the visible methods for demonstrating customer focus are shown in Fig. 3.5. The methods could be applied by top management during discussions with senior management personnel for demonstrating customer focus are:

(i) Monitoring the effectiveness of interaction between support processes and production processes for meeting customer requirements.
(ii) Providing people and infrastructure for meeting the requirements of products and customers.
(iii) Addressing customer needs when cost and time reduction measures are initiated for improving the efficiency of processes.

The discussions could be recorded and the records inspire senior management personnel for creating customer focus in their respective areas. The management

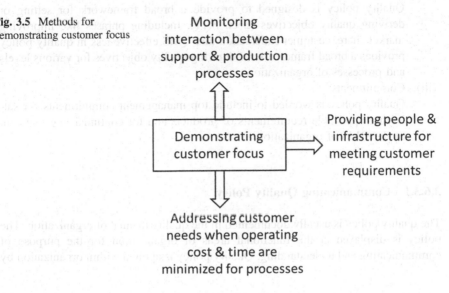

Fig. 3.5 Methods for demonstrating customer focus

personnel demonstrate their commitment with respect to customer focus by ensuring that:

(i)   Customer and applicable statutory and regulatory requirements are determined, understood and consistently met. The requirements are integrated with appropriate QMS procedures, implemented and monitored.
(ii) The risks and opportunities that could affect conformity of products and the ability to enhance customer satisfaction are determined and integrated with QMS procedures as explained in Sects. 3.7–3.10.
(iii) The focus on enhancing customer satisfaction is maintained. Conducting periodical reviews for delivering products with consistent quality as per schedule and prompt customer communications to enhance customer satisfaction are ensured.

3.6.3 Establishing and Communicating Quality Policy

Quality policy is a formal statement of top management expressing their intentions and directions for organization. Top management establishes the quality policy of organization considering the requirements explained below:

(i)   Appropriate to organization needs:
      Quality policy addresses appropriately the purpose, context and the strategic direction of organization (Sect. 3.2). Phrases like compliance to legal requirements and updating technology to ensure competitive advantage could be considered for establishing quality policy.
(ii) Framework for quality objectives:
      Quality policy is designed to provide a broad framework for setting or deriving quality objectives (Sect. 3.11). Including phrases like improving market share, customer satisfaction and QMS effectiveness in quality policy provides a broad framework for deriving quality objectives for various levels and processes of organization.
(iii) Commitments:
      Quality policy is worded to include top management commitments for satisfying applicable requirements of products and for continual improvement of the QMS of organization.

3.6.3.1 Communicating Quality Policy

The quality policy is usually documented in the quality manual of organization. The policy is displayed at the functional areas of organization for the purpose of communicating and understanding. Quality policy is applied within organization by implementing the quality objectives for various functions. Copies of quality manual are distributed to the relevant interested parties within organization, communicating the quality policy.

3.6.4 Organizational Roles, Responsibilities and Authorities

Roles are positions for individuals. Managerial tasks are associated with the positions. Top management assigns roles to individuals (process owners) for managing a process or a group of processes in organization. Responsibilities are the expected results from process owners. Authorities are the administrative powers for the process owners to manage the processes. Responsibilities and authorities drive process owners for achieving planned results. Top management defines the responsibilities and authorities of process owners appropriate to their roles. The responsibilities of process owners are:

  (i) Ensuring that the QMS documentation conforms to the requirements of ISO 9001. Process owners usually approve their quality procedures.
 (ii) Ensuring that the processes are delivering their intended (planned) outputs as relevant QMS documents.
(iii) Reporting the performance of the assigned processes and opportunities for improvement (Sect. 6.7) to top management as inputs for reviews (Sect. 6.6.1).
 (iv) Ensuring customer focus (Sect. 3.6.2) in the functional areas of process owners.
  (v) Ensuring that the integrity of the QMS documentation is maintained when changes to the documentation are planned and implemented by process owners.

Usually, the roles, responsibilities and authorities assigned by top management to process owners are documented in quality manual and the manual is distributed. They are also displayed in functional areas for communication and understanding within organization. Top management could assign the roles, responsibilities and authorities for one of the senior management personnel of organization for monitoring the implementation of the QMS of organization.

3.7 Planning

Planning is an intellectual process of thinking before acting [4]. It is essential to achieve the intended benefits of QMS explained in Sect. 3.1. Planning needs to be done by everyone in organization, but it is significant for process owners considering the impacts on achieving the planned outputs of QMS processes. The requirements of planning as per ISO 9001 are:

(i) Actions to address risks and opportunities:

- Understanding risks and opportunities
- Determining risks and opportunities
- Planning the actions to address risks and opportunities
- Examples of planning the actions to address risks and opportunities, integrating the actions with QMS processes and evaluating the effectiveness of the actions for three QMS processes.

(ii) Quality objectives:

- Requirements for establishing quality objectives
- Planning to achieve quality objectives

(iii) Planning of changes

3.7.1 Understanding Risks and Opportunities

Risk is the probability and severity of failure to achieve the desired outputs (results) of processes. Probability of the failure is the likelihood of risk occurrence and it is classified as impossible-Low-Medium-High-Certain qualitatively. Severity represents the potential impact of failure in achieving the desired outputs and it is classified as Minor-Major-Critical-Catastrophe. Most risk management guidelines recognize at least four risk response strategy types, namely, avoid, transfer, mitigate and accept [5]. ISO 31000 suggests risk treatment by avoiding, treating risk sources, modifying likelihood, changing sequences or sharing the elements of risk and the residual level of risk after treatment should be within risk appetite i.e. within acceptable level.

Opportunity is a favorable opening to achieve the desired results of processes overcoming the risks in the processes. Innovation and knowledge serve as sources to identify opportune actions. Some of the sources to identify the actions are learning from past experience, adoption of emerging technologies or practices and learning from competitors. The opportune actions address risk management strategies (avoid, transfer, mitigate and accept) appropriately considering the probability and severity of risk. The finalized actions should be able to achieve the desired results of processes almost with certainty and implementable within the constraints of organization. Examples are provided for understanding the methods of addressing risks and opportunities for QMS processes (Sects. 3.8–3.10). The general procedure for addressing risks and opportunities in processes is shown in Fig. 3.6.

Fig. 3.6 General procedure for addressing risks and opportunities

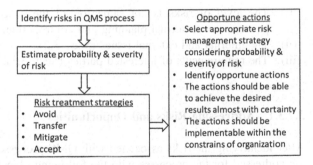

3.7.2 Practical Considerations

Recognizing the existence of risks that impede the achievement of desired results of processes is fundamental to identify the actions to address risks and opportunities of QMS processes. Downplaying the risks increases the probability failures in achieving the planned results of the processes and closes the opportunities for improving the processes. Risk-based thinking builds strong knowledge base, establishes proactive culture for improvement and improves customer satisfaction [6]. Organizing brainstorming sessions with relevant process owners is appropriate for identifying the actions. Group discussions synergize the innovative abilities and knowledge of individuals to identify practical and cost effective actions within the constraints of organization.

Corrective actions are identified after the occurrence of failures in processes and products by performing analysis to identify root causes of the failures. Preventive actions could be identified for processes based on the learning from corrective actions. However, identifying preventive actions by addressing the risks and opportunities of the processes is systematic and realistic.

Failure mode and effects analysis (FMEA) systematically identifies preventive actions for the postulated failure modes of processes and products. FMEA is therefore more suitable for addressing risks and opportunities associated with QMS processes. The application of FMEA technique is explained for electronic products [7] and the technique could be extended for the processes.

3.7.3 Determining Risks and Opportunities

Planning of QMS refers to determining the processes required for the QMS and defining sequence, interactions and other related characteristics of the processes. Risks are associated with the processes in achieving the intended outputs from the processes and they are considered in planning the QMS of organization. Risks associated with QMS processes are identified considering the following:

(i)   The level of risks is not same for all the processes of QMS and some processes need formal planning and controls than others [6].
(ii) The internal and external issues (Sect. 3.2).
(iii) The requirements of interested parties (Sect. 3.3).

3.7.3.1 Addressing Risks and Opportunities

After identifying the risks associated with QMS processes, risks and opportunities are addressed for the processes with the following objectives:

(i)   Assuring that the QMS achieves the intended results.
(ii) Enhancing desirable effects through opportune decisions.
(iii) Preventing or reducing undesired effects by treating risks.
(iv) Achieving improvements for QMS processes.

Addressing risks and opportunities is integral part of decision making in organization to prevent failures in processes. Senior management personnel discuss with relevant process owners periodically to review customer orders. Risks and opportunities are intuitively addressed during the group discussion with the objective of ensuring conformance to product and customer requirements. The planned actions (strategies for managing risks and opportunities) are recorded. The actions are implemented and evaluated for their effectiveness in subsequent reviews. The course of further actions is decided in the reviews. Similar approach is required for planning actions to address risks and opportunities in QMS processes.

3.7.4 Planning Actions to Address Risks and Opportunities

ISO 9001 specifies that actions to address the risks and opportunities of QMS processes should be identified and the actions should be formally integrated with QMS processes to ensure conformity of products and services. Planning the actions for the processes requires:

(i)   Identify internal and external issues and the requirements of interested parties relevant to the risk.
(ii) Identifying the causes of risks of QMS process.
(iii) Identifying the actions arising from internal and external opportunities considering the risk response strategies associated with the opportunities.
(iv) Integrating the finalized opportune actions with the documentation of the QMS process.
(v)   Monitoring the actions.
(vi) Analyzing data from monitoring.

3.7.4.1 Managing Risks

Actions taken to address risks and opportunities are proportionate to the potential impact (severity) on the conformity of products. The management strategy is to avoid risks that have high probability of occurrence or that could cause severe impact on the desired outputs of processes. The strategy is to accept risks that have low probability of occurrence and that could cause minor impact on the outputs of processes.

The methods of addressing risks and opportunities and the actions taken are explained for three QMS processes (Sects. 3.8–3.10) for understanding and developing the documented procedures. Additional information to improve the effectiveness of planning the actions to address risks and opportunities is presented in Chap. 9 using the ideas of Indian classic literature.

3.7.4.2 Maintaining Records

Quality records are maintained for integrating the planned actions to address risks and opportunities in the documents of the processes. Monitoring and measurements are performed to evaluate whether the planned actions are implemented effectively. The methods monitoring and measuring are indicated in the examples for three QMS processes and in Sect. 6.2.4.2. The results of evaluation are inputs for management reviews for evaluating the effectiveness of the actions taken. The records serve as a source of organizational knowledge for addressing risks and opportunities.

3.8 Example-1: Risks and Opportunities in Design and Development

Uncertainties (risks) generally exist in the design and development product to satisfy performance, delivery, cost and other requirements of customers. Technology development in markets is an external issue for organization. The issue becomes relevant to organization when customers demand state of the art products and the organization accepts to design and develop the products. Managing risks and opportunities in the design and development of products for the technology issue is explained with an example.

Assume that organization is required to design and develop a complex communication system for a customer. The electronic units that are required for the design and development of the system are identified. Planning is done to organize the units for completing the design and development of the system cost effectively, satisfying performance, reliability and delivery requirements.

During planning, one of the electronic units, Unit-ABT, is identified to have risks, impeding the achievement of the delivery requirements of the system. Brain storming session with designers and other relevant personnel is organized for identifying the causes of the risks associated with the Unit-ABT. The actions addressing opportunities are also identified during the session to manage the risk.

3.8.1 Risks and Their Causes

All possible causes of the risk associated with the Unit-ABT and the causes of the risks are identified and documented. Inadequate knowledge for the design and development of Unit-ABT is the primary cause of the risk. To keep it simple, only the primary cause, severity (potential impact) and probability of risk are shown in Table 3.1.

3.8.2 Identifying the Actions to Address Opportunities

The internal and external sources are utilized to identify the actions to address opportunities (or simply opportunities) for managing the risk associated with the Unit-ABT. Some of the identified opportunities from internal and external sources with their linkage to risk response strategies are shown in Tables 3.2 and 3.3 respectively for understanding.

Table 3.1 Details of risks in Unit-3

Risk	Primary cause of the risk	Severity of risk	Probability of risk
Delay in system delivery	No previous experience in the design and development the unit	Critical	Medium

Identification: Unit-ABT

Table 3.2 Opportunities from internal sources with risk response strategy

Actions arising from internal opportunity	Risk response strategy
Past internal records are examined to identify the opportunities that were successful in managing similar risks	Mitigating risk
Formation of expert design team	Mitigating risk
Access to leading digital libraries around the world to use emerging technologies	Mitigating risk

Table 3.3 Opportunities from external sources with risk response strategy

Actions arising from external opportunity	Risk response strategy
Hiring related experts	Avoiding risk
Learning from competitors	Mitigating risk
Receiving the Unit-ABT as customer property	Transferring risk

3.8.3 Integrating the Actions with QMS Process

As the potential impact (severity) of risk associated with the Unit-ABT is critical to the communication system, one or a combination of the identified actions that would avoid the risk should be preferred considering the constraints of organization. The finalized opportune actions addressing the risk associated with the Unit-ABT are documented. The actions could be linked with design and development planning of the communication system.

Linking the actions with design and development planning facilitates the implementation of the action, monitoring and measuring the action for achieving the planned results and the generation of required documentary evidences. A simple measurement for the actions addressing the risk of Unit-ABT would be to measure the time for realizing the unit. The progress of the Unit-ABT is reviewed by project team. The measured time could be compared with the planned time to evaluate the effectiveness of the actions. The effectiveness of the actions taken to address the risks and opportunities for the design and development process is evaluated in management reviews.

3.9 Example-2: Risks and Opportunities in Customer Enquiries

Enquiries for products are received from customers and quotations are sent to the customers, committing delivery schedule for the quantity of the products. Quotations imply that organization is committed to deliver the quantity of products as per schedule. Uncertainties (risks) might exist in some of the enquiries of customers. Actions to address the risks and opportunities associated with the enquiries need to be planned before sending quotation to customers. Managing risks and opportunities for customer enquiries having risks is explained with an example.

Consider an organization with its scope defined as designing, manufacturing and delivering microwave components to customers. It is assumed that the organization has received an enquiry from customer for delivering large quantity of standard microwave component with delivery schedule, not compatible to the quantity. As some of the facilities of organization are inadequate to manufacture large quantity of the component within the specified delivery schedule, risks are associated in accepting the delivery schedule. However, considering the expectations of

Table 3.4 Details of risks in accepting customer enquiry

Risk	Primary causes of the risk	Severity of risk	Probability of risk
Inability to manufacture large quantity as per delivery schedule	Inadequate operators	Minor	Medium
	Inadequate external providers	Major	Medium
	Inadequate test equipment	Critical	Medium

interested parties (employees and investors), the enquiry is treated as opportunity. The actions addressing the opportunity are planned to manage the risk before sending the quotation for the customer enquiry.

3.9.1   Risks and Their Causes

All possible causes of the risk associated with the acceptance of the enquiry are identified and documented. To keep it simple, only the primary causes, severity (potential impact) and probability of risk are shown in Table 3.4. Identifying the actions addressing opportunities to manage the risk is presented.

3.9.2   Identifying the Actions to Address Opportunities

Both internal and external sources are utilized to identify the opportunities for managing the risk associated with the acceptance of the enquiry. Opportunities from internal and external sources are explored to manage the risks. Examples of opportune actions are identifying external providers, transferring or hiring operators and renting or buying test equipment. The identified opportune actions should avoid (risk response strategy) the risk, considering the needs of interested parties. The finalized opportune actions are documented.

3.9.3   Integrating the Actions with QMS Process

The finalized opportune actions could be linked with operational planning and control for production processes. The arrangement facilitates the implementation of the actions, measuring the effectiveness of the actions and the generation of required documentary evidences. The progress of the customer order is reviewed by production management team. On-time delivery is usually a quality objective measuring the effectiveness of operational planning and control. The same arrangement could be used for measuring the effectiveness of the opportune actions for executing the customer order. The actions and the effectiveness of the actions to address the risks and opportunities in accepting the short delivery schedule for the microwave component are evaluated in management reviews.

3.10 Example-3: Risks and Opportunities in Production Processes

Production operations are a set of processes with defined sequences and interactions of the processes for realizing product. The processes are implemented under controlled conditions, which are in the form of documented information i.e. product related engineering documents. Uncertainties (risks) might exist in production processes in achieving the planned results of the processes consistently. Examples of the risks associated with production processes are:

(i)   Operator dependent processes.
(ii) Improper jigs and fixtures for performing operations.
(iii) Performance measurements are not repeatable.

Assume that the severity of the risks is minor and the probability of the risks is medium. The risks might progress to become failures in production processes, affecting product conformance and delivery. The methods of managing risks and opportunities in production processes are explained.

3.10.1 Opportune Actions and Integrating Them with QMS Process

Actions to address opportunities are identified for managing operator dependent production process. Utilizing internal sources would be adequate to identify the opportunities. The identified opportunities should avoid (risk response strategy) the risk.

The risk in operator dependent production process indicates that defects might occur if the expert operator of the process is not available for any reason to perform the operation. The cause of the risk is that the finer details performed by the expert operator are not reflected in the work instruction document of the process. The opportunity is to understand the finer details performed by the expert operator and amend the work instruction document of the process to include the finer details. The innovative efforts of expert operators enhance organizational knowledge and hence the contribution of the operators should be recognized when work instruction documents of production processes are amended. Alternatively, automation for the operations could also be considered, eliminating the variations in processes.

Identifying actions to address risks and opportunities are explained for three QMS processes. It could be observed that the level of risks is not same for the processes. The process, Design and development of products, requires higher level of planning and controls than others.

3.11 Quality Objectives

Quality objectives are decided for improving the effectiveness of QMS processes and they should be measurable as per the requirements of ISO 9001. They are established at relevant functions and levels as appropriate for support and production processes of QMS. Examples of quality objectives are:

   (i)   Design changes initiated by the users of engineering documents.
  (ii)   Rejection level in the assembly processes of products.
 (iii)   Statistical variations for the critical characteristics of products.
 (iv)   Trends in customer complaints for products.
  (v)   Level of customer satisfaction.
 (vi)   On-time delivery for production planning process.

It is convenient to derive quality objectives for the processes when characterizing the processes as explained in Sect. 2.4. The requirements that should be considered for deciding the quality objectives of QMS processes are explained.

3.11.1 Requirements

When identifying and deciding quality objectives, the requirements of the objectives are:

  (i)   Consistent with the quality policy of organization:
      Quality objectives are the means to demonstrate the implementation of quality policy. Hence, they are derived from quality policy and linked to the appropriate functions, levels and processes of QMS.
 (ii)   Measurable:
      Appropriate quantitative metrics are developed for measuring quality objectives. The measurements enable evaluating the effectiveness of processes. Number of design changes initiated by the users of engineering documents and percentage of rejections in the assembly processes of products are examples of measurable quality objectives.

(iii) Take into account applicable requirements:
Quality objectives are identified considering the requirements of customers and relevance to the processes of QMS. Statutory and regulatory requirements of customers are also considered.
(iv) Relevant to product conformity and to enhance customer satisfaction:
Quality objectives are decided for operational planning, customer communication, design and development, production processes and other processes that are relevant to product conformity and to enhance customer satisfaction.
(v) Monitored:
The levels of achieving quality objectives are monitored and measured.
(vi) Communicated:
The measured values of quality objectives are analyzed and the results of the analysis are made available for management reviews for evaluation.
(vii) Updated as appropriate:
Based on the outputs of management review, quality objectives are updated as appropriate.

Process owners propose quality objectives of processes and products in most organizations. Top management reviews and finalizes the proposed quality objectives after ensuring that the objectives are measurable and consistent with the quality policy of organization. The finalized list of quality objectives are communicated to the process owners for implementation. The list serves as documented information (quality records) on the quality objectives.

3.11.2 Planning to Achieve Quality Objectives

Process owners plan the methods of achieving their quality objectives. The following are determined as part of planning:

(i) What needs to be done:
Quality objective is broken down to list of actions for implementation. The actions include monitoring and measurements relevant for the quality objective.
(ii) What resources are required:
The resources required for implementing the actions are identified.
(iii) Who will be responsible:
Process owners are responsible for implementing the actions and achieving the objective of the process.
(iv) When it will be completed:
Usually, quality objectives are the annual objectives of organization. However, planning could decide a different time frame to achieve the objective.

(v)   How the results will be evaluated:
      The results of monitoring and measurements are used to express quality
      objective quantitatively. Average or weighted average or mathematical
      expressions could be used to analyze the measured values. The results of the
      analysis are used to evaluate the level of achievement of the quality objec-
      tive. The methods of evaluating the results are decided during planning.
      Additional information for evaluating quality objectives are provided in
      Sect. 6.2.

3.11.3   Planning Changes

Changes to QMS documents might arise due to corrective actions, improvement
actions, revising the scope of organization, etc. The changes are carried out in the
planned manner, described for establishing the QMS (Sect. 3.5). Organization
considers the following before approving the changes to the processes of QMS:

  (i)   The purpose of the changes and their potential consequences:
        The source (audit findings, corrective actions, etc.) and the purpose (reasons)
        of changes to the QMS processes are documented. The expected benefits of
        the changes are also recorded.
 (ii)   The integrity of QMS:
        The proposed changes are examined and ensured that the changes comply
        with the requirements of the QMS of organization.
(iii)   The availability of resources:
        Most of the changes to production related processes would require resources,
        which might be simple fixtures, measuring devices, improving competence
        of personnel, etc. The availability of resources is ensured before effecting
        changes to QMS documents.
(iv)   The allocation or reallocation of responsibilities and authorities:
        Changes initiated by top management might require the allocation or real-
        location of responsibilities and authorities for personnel. The changes are
        reflected appropriately in the related documents of QMS.

References

1. ISO 9001:2015, How to use it, iso.org, ISBN 978-92-67-10640-3
2. ISO 9000:2005 Quality management systems—fundamentals and vocabulary
3. Baron Robert (1983) Behavior in organization: understanding and managing the human side of
   work. Allyn and Bacon Inc, USA
4. Robbins S, Coulter M (2006) Management. Prentice Hall

5. Hillson D (2001) Effective strategies for exploiting opportunities. In Proceedings of the project management institute annual seminars and symposium, Nashville, Tenn., USA
6. Based R, Thinking in ISO 9001:2015, ISO/TC-176/SC2/N1269
7. Natarajan D (2015) Reliable design of electronic equipment: an engineering guide. Springer

Chapter 4
Support Processes

Abstract In addition to the commitment of top management and planning, support processes are needed for implementing QMS processes. Resources, monitoring and measuring resources, organizational knowledge and documented information are required for the operation of QMS processes. The requirements of the support processes, competence and communication are presented.

4.1 Introduction

In addition to the management processes described in Chap. 3, support processes are necessary for implementing operational and other processes of organization. People, infrastructure and suitable environment for the operation of the processes are required. Monitoring and measuring resources are required to support the design, development and production processes of products. Training might be required to ensure the necessary competence of persons performing the operations of processes. QMS documents including product related engineering documents needs to be controlled for ensuring the planned outputs of processes. The requirements of support processes are presented.

4.2 Resources

Resources for the establishment, implementation, maintenance and continual improvement of the QMS of organization are determined and provided. The categories of resources are:

(i) People, infrastructure and environment.
(ii) Monitoring and measuring resources.
(iii) Organizational knowledge.

If the internal resources are inadequate in terms of capabilities or if constraints exist for organizing them internally, the resource needs are obtained from external providers. Calibration services and hiring experts are examples of resource needs obtained from external providers.

4.2.1 People, Infrastructure and Environment

Persons are required for the effective implementation of the QMS of organization and for the operation and control of the processes of QMS. Top management plans the requirements of personnel and provides them. Additional requirements are usually identified by process owners and top management provides the resources after review.

4.2.1.1 Infrastructure

Infrastructure facilities are necessary for the operation of processes and to achieve conformity of products. Organizations need buildings with associated utilities, ERP software for planning, manufacturing equipment for the operation of processes, test equipment to verify conformity of products, internet services and other unique infrastructure facilities. The infrastructure facilities are determined, provided and maintained by organizations. Most organizations document the infrastructure facilities, including monitoring and measuring resources with the brief capabilities of the resources to support maintenance.

4.2.1.2 Environment

Environment refers to the surrounding conditions for the operation of processes and to achieve conformity of products. Some of the processes of QMS require controlling the physical factors of environment for their operation. Semiconductor assembly processes require dust, temperature and humidity controlled environment. Testing of acoustic devices requires noise controlled environment. Appropriate social and psychological environments are also maintained for carrying out processes and they also contribute for achieving product conformity.

Fostering team work and avoiding conflicts are examples of maintaining positive social environment for the operation of processes. Calm environment would be desired for the design and development of products. Maximizing computerization for the operation of processes provides stress-reducing environment for personnel psychologically. Organization determines, provides and maintains appropriate combination of environment considering physical, social and psychological factors for the operation of processes and to achieve conformity of products.

4.3 Monitoring and Measuring Resources

Organizations use monitoring and measuring resources for design, development and production processes. The resources are generally electronic equipment and mechanical measuring devices. In-house designed Accept-Reject test fixtures are also part of the resources. Examples of mechanical measuring devices are Micrometers, Hardness testers and Profile projectors. Examples of electronic test equipment are Signal generators, Oscilloscopes and Network analyzers.

Monitoring equipment is used for checking the status of process parameters for control purposes to ensure outputs with consistent quality. For example, temperature indicators are used for monitoring solder bath for PCB assembly process. Nanoparticles measuring equipment is used for monitoring the dust level of clean rooms for semiconductor manufacturing processes. Monitoring equipment actually measures the parameters of process control variables.

Measuring equipment is used for quantifying the parameters of in-process and finished products for verifying conformity of products to requirements. Mechanical measuring devices and electronic test equipment are used for quantifying the dimensional, physical and electrical parameters of products. The requirements for determining and providing resources for monitoring and measurement to ensure valid and reliable results are explained.

4.3.1 Determining and Providing Resources for Measurement

Engineering documents specify the monitoring and measurement parameters of products and processes. For example, engineering drawings specify the dimensions of parts with tolerance. Finished product documents specify the electrical characteristics of the products. The engineering documents are controlled by design team and they are the basis to determine monitoring and measuring parameters for processes and products.

4.3.1.1 Providing Resources for Measurement

Suitable measuring resources are selected for the determined parameters. Resolution and accuracy are the two characteristics that should be considered for selecting resources for measuring the parameters of products and processes. Least count instead of resolution is used for selecting non-electronic mechanical measuring devices.

Accuracy is the value of uncertainty in measurement parameter and the value of the uncertainty is expressed by the tolerance of the parameter. The recommended values of resolution and accuracy for selecting resources are [1]:

(i)   Resolution: Better than 1/10 of tolerance.
(ii) Accuracy: Better than 1/3 of tolerance.

Example 1 Specified external diameter of a machined part: (10.40 ± 0.03) mm. The nominal diameter is 10.40 mm and the tolerance on the nominal diameter is ±0.03 mm. The external micrometer selected for measuring the diameter of the part should have a resolution of 0.003 mm or better and its accuracy should be ±0.01 mm or better.

Example 2 Specified output voltage of DC Power Unit: (5.0 ± 0.1) V. The nominal output voltage is 5.0 V and the tolerance on the nominal voltage is ±0.1 V. The equipment selected for measuring the output voltage should have a resolution of 0.01 V or better and its accuracy should be ±0.03 V or better.

4.3.1.2   Practical Considerations

Digital signal processing with ICs has revolutionized both mechanical measuring devices and electronic test equipment. Most of the hand held mechanical measuring devices like calipers and micrometers are available with digital read-outs. High resolution, high accuracy, light weight and smaller size are the characteristics of electronic equipment for most measurement applications. Hence, it is adequate to focus on the monitoring and measurement characteristics having very low tolerance for the selection of equipment.

4.3.1.3   Linking Measurement Parameters with Resources

Product related work instructions are prepared by design team for verification of externally provided items, inspection during and after production processes and other measurement processes. The work instruction documents are used to link the determined measurement parameters with suitable measuring resources. Measuring resources, having appropriate resolution and accuracy, are specified satisfying the tolerance requirements of parameters. Table 4.1 shows an example of work instruction document linking measurement parameters with measuring resource for the verification of externally provided parts.

Table 4.1 Linking measurement parameters with measuring resource

Work Instructions			
Verification of externally provided parts			
Engineering drawing: MC547		Part: MICCY	
Parameter	Resource	Measured value	Remarks
(9.80 ± 0.03) mm	HIG123		

4.3.2   Fitness of Measuring Resources

Measuring resources are used for verifying product characteristics, process performance and control parameters. The resources need to be maintained for ensuring continued fitness for their purpose. FDA regulation 21CFR820.72 specifies the requirements for ensuring the accuracy and fitness of measuring resources. The requirements are shown in Fig. 4.1 and the requirements are [2]:

(i) Measuring resources should be routinely calibrated, inspected, checked, and maintained.
(ii) The resources should be handled, preserved and stored appropriately.

4.3.2.1   Calibration, Inspection and Checking

One or a combination of the three methods (calibration, inspection and checking) is used to ensure the accuracy and fitness of measuring resources. Inspection is usually integrated with calibration and checking. Visual inspection of calipers for damages to internal and external jaws and zero errors is part of calibration or checking. Visual inspection for damages is used for ensuring the fitness of steel rules. Calibration and checking are explained and illustrated with examples in subsequent sections.

4.3.2.2   Maintenance

Maintenance refers to preventive and breakdown maintenance for measuring resources. Equipment used for monitoring safety and product critical processes requires preventive maintenance. Applicable method of calibration or checking is applied for ensuring the fitness of the measuring resources after preventive and breakdown maintenance.

Fig. 4.1 Ensuring the accuracy and fitness of measuring resources

4.3.2.3 Quality Procedures and Records

Organization maintains documented information (quality procedures) describing appropriate methods for ensuring the accuracy and fitness of monitoring and measuring resources. Documented information (quality records) is also retained to provide evidences for the results of implementing the methods.

4.3.3 Measurement Traceability

The fitness of measuring resource is verified by comparing its performance with the performance of measurement standards having higher accuracy and resolution than the measuring source. The lowest level of measurement standard is popularly known as working standard. The fitness of the working standards is also verified by comparing it with next higher level of standards (secondary standards) and so on. A chain of measurement comparisons could be visualized in the efforts to ensure valid and reliable results on products and processes, leading to the definition of traceability in measurements.

4.3.3.1 Definition of Traceability

Traceability is defined as "the property of a result of a measurement whereby it can be related to appropriate standards, generally national or international standards, through an unbroken chain of comparisons" [3]. In other words, the calibration of measuring device or equipment should be traceable to national measurement standards. The simplified calibration traceability diagram with three levels is shown in Fig. 4.2 and the levels are:

Fig. 4.2 Calibration traceability chain for three levels

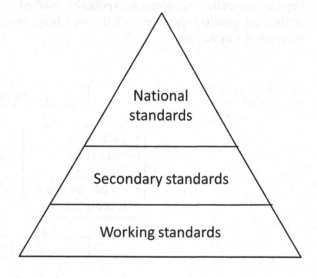

(i) Ensuring the compliance status of the measuring resources of organization using working standards.
(ii) Establishing the compliance status of working standards using secondary standards.
(iii) Establishing the compliance status of secondary standards using national standards.

4.3.3.2   Measuring Resources Requiring Traceability

Measurement traceability is ensured for the following categories of standards and measuring resources of organization to provide confidence in the validity of measurement results:

(i) Mechanical measuring devices and equipment.
(ii) Electronic equipment.
(iii) Working and secondary standards.
(iv) The instruments mandated by regulatory bodies such as Occupational Safety and Health Administration [3].

4.3.3.3   Ensuring Measurement Traceability

ISO 9001 provides options (methods) to ensure the traceability of measuring resources considering the availability of measurement standards. The methods to ensure measurement traceability are:

(i) Calibration against measurement standards.
(ii) Verification against measurement standards.
(iii) Calibration or verification as per established procedures when measurement standards do not exist.

Implementation of the methods for the measurement resources of organization is explained. The methods are applied at specified intervals or prior to use. Suggestions are provided for deciding the intervals.

4.3.4   Calibration Against Measurement Standards

Calibration should be carried out at laboratories accredited by national laboratory or at in house as per documented method by competent staff using reference standards with measurement traceability to national laboratory [4]. National and international standards exist describing the method for performing calibration against measurement standards. The standards specify the competence of personnel, infrastructure,

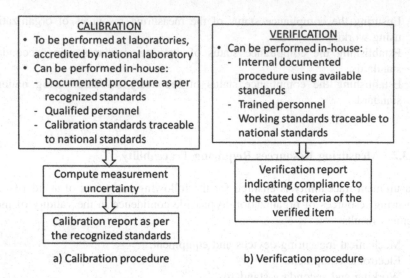

Fig. 4.3 Calibration and verification of measuring resources

calibration environment, accuracy and resolution requirements of measurement standards, computation of measurement uncertainty and preparation of calibration reports.

Large organizations maintain in-house calibration laboratories conforming to the accreditation requirements of national laboratories. Working and secondary standards are maintained by the laboratories with traceability to national standards. Organization could also use the services of external accredited laboratories for calibrating their measuring resources. The general procedure for calibration is shown in Fig. 4.3a.

4.3.4.1   Calibration Intervals

Calibration interval for equipment is generally one year. It could be increased if the performance of the equipment is much better than requirement [5]. Equipment performance is considered better if the accuracy and resolution of the equipment is much higher than the tolerance and resolution requirements of measurements. Drift, tendency to wear, recorded history of maintenance and handling and storage arrangements are also considered for deciding calibration interval [1]. Measuring equipment is re-calibrated considering the nature of maintenance, affecting the reliability of measurement. It might so happen that some projects might be closed for various reasons in organizations. The equipment associated with the projects is suitably identified for not being calibrated. When required, it is calibrated prior to use.

4.3.4.2 Constraints

It is not feasible for most organizations to calibrate all the measuring resources at external accredited laboratories or establish in-house accredited laboratories conforming to the requirements of national laboratories due to cost and time constraints. Alternatively, in-house verification facility could be established for verifying the fitness of measuring resources using measurement standards. The measurement standards are usually working standards.

4.3.5 Verification Against Measurement Standards

Verification is conformation that specified requirements have been fulfilled through observation, measurement, test or other means [6]. Verification is checking the performance of measuring equipment against working standards. Checking is carried out in-house by competent staff using appropriately calibrated equipment to indicate compliance or otherwise with stated criteria [4].

Accreditation by national laboratory is not needed for performing in-house verification activities. Unlike calibration, verification of measuring resources does not require calculation of measurement uncertainty [7]. Verification indicates compliance or otherwise with stated criteria i.e. whether the measuring resources are fit or not for providing confidence in the validity of measuring results.

Working standards, used for verification activities, are calibrated at national accredited laboratories to establish traceability to national standards. Quality assurance personnel could be assigned with the roles, responsibilities and authorities and trained for performing in-house verification activities. Generally, organization uses both calibration and verification for providing confidence in the validity of measurement results. The general procedure for calibration is shown in Fig. 4.3b.

4.3.5.1 Services of External Providers

The services of the suppliers of electronic equipment or external providers could be obtained for performing verification of measuring equipment. The suppliers and the external providers should also use appropriately calibrated working standards for checking the fitness of the equipment. Documented procedures and records are maintained for carrying out verification on the measuring resources of organization.

4.3.5.2 Examples for Verification Using Working Standards

Working standards could be used for ensuring the fitness of mechanical measuring devices and electronic equipment. Calibrated slip gauges are used to ensure the fitness of in-house mechanical measuring devices. Calibrated standard resistors are used to ensure the fitness of in-house resistance measuring equipment. Calibrated standard multi-meter is used to ensure the fitness of other multi-meters and DC power supplies. Similar procedure is applied for ensuring the fitness of software and video based equipment such as Coordinate Measuring Machine (CMM). The accuracy and resolution requirements of working standards should be considered for selecting the standards.

4.3.5.3 Accuracy and Resolution of Working Standards

The accuracy and resolution of working standard should be higher than those of equipment to be verified. Accuracy of working standard is expressed as Accuracy ratio or Test accuracy ratio. It is the ratio of the tolerance of the parameter of equipment for measuring equipment to the tolerance of the parameter of working standard. The accuracy of working standards should be better than four times that of equipment to be calibrated i.e. the accuracy ratio should be 4:1 or better [3, 8].

Consider the same example 1 in Sect. 4.3.1.1. The specified external diameter of a machined part: (10.40 ± 0.03) mm. External micrometer having accuracy of ± 0.01 mm or better is required for measuring the dimensions of the part. Assume that working standard satisfying the accuracy ratio 4:1 is required for checking the fitness of the micrometer.

$$Accuracy\ Ratio = \frac{Tolerance\ of\ external\ micrometer}{Tolerance\ of\ working\ standard}$$

$$4 = \frac{0.01}{Tolerance\ of\ working\ standard}$$

$$Tolerance\ of\ working\ standard = 0.01/4 = 0.0025$$

Working standard with accuracy of ± 0.002 mm or better is required for ensuring the fitness of the external micrometer. The resolution of working standard should be better than two times that of equipment to be calibrated [1]. Similar computations could be made for selecting working standards with adequate accuracy and resolution to check the fitness of electronic equipment. The factors that decide the verification interval for measuring resources are same as those for calibration interval.

4.3.6 Non-existent of Measurement Standards

Measurement standards do not exist for many parameters of process related measurements. In such cases, valid calibration is performed against reference material, certified reference materials or consensus industry standards [1]. The fundamental laws of physics and chemistry are the basis of calibration or verification when measurement standards do not exist. Examples for calibrating and verifying process related monitoring equipment are provided.

Appropriate boiling liquid baths with corrections to barometric atmospheric pressure are used for checking the fitness of mercury thermometers. Pressure gauges can be calibrated using a dead weight tester, where pressure is first created through the piston and cylinder arrangement and then the same is balanced against calibrated weights [1]. In-house designed Accept-Reject test fixtures for the production processes of products are inspected visually and calibrated or verified for fitness using appropriate equipment. pH meters are verified for pH accuracy and linearity using commercially-prepared buffers or standard buffers (as specified in a pharmacopoeia) according to the manufacturer's instructions daily in use [9]. The calibration and verification procedures are maintained by organizations. References [1, 4, 9] provide practical examples for performing calibration and verification on process monitoring equipment with intervals for the activities.

4.3.7 Calibration with Lower Accuracy Ratio Standards

Accuracy ratio of 4:1 might not be feasible for measurement standards due to technological advancement in instrumentation. Guard-banding, as proposed in ISO 17025 and in other technical literature offers a statistical technique for addressing lower accuracy ratio for calibration standards. Its application is explained for calibrating Digital Multi-meter with a calibration standard having an accuracy ratio, 2.4:1 [10]. Standardized Normal distribution is used for computing the double sided guard-band tolerance limits for the equipment under calibration (EUC) at 90% confidence level. The guard band tolerance limits decide pass or fail criteria for the EUC. The technique provides 95% confidence level for the parameters of EUC with single sided guard-band tolerance limit.

4.3.8 Calibration Status and Safeguarding

Calibration or verification status of equipment is usually identified by stickers or color coded on measuring resources. Linking the serial number of measuring resource with its calibration or verification report is also acceptable. Where possible, equipment is safeguarded from inadvertent adjustments by users that would invalidate the measurement results. Equipment is protected from damage and deterioration during handling, maintenance and storage. Standard operating environmental conditions are maintained for measurement standards.

When measuring resource is suspect or found to be unfit for providing valid measurement results, the calibration or verification of the equipment is checked or the measurements are repeated in similar calibrated equipment. If the equipment is confirmed to be faulty, appropriate actions such as re-inspecting the products in shop floor and in stores are taken. If required, corrections might be needed for the products delivered.

4.4   Organizational Knowledge

Organization needs knowledge for the satisfactory operation of processes to ensure the achievement of conformity of products and services. The internal sources for the knowledge are:

   (i) Capabilities and application of measuring resources:
       Understanding the capabilities and application of measuring resources enables to use the resources efficiently and effectively for the production processes of products.
  (ii) Improving competency of personnel:
       Competency of personnel is a source of organizational knowledge. Appropriate education and training could be organized for persons managing product design and production processes.
 (iii) Lessons learned:
       Knowledge sources exist in organization. Lessons learned from the root cause analysis of process and product failures and the successful management risks in processes are examples of the sources. Problems solved by operators under suggestion award scheme should also be treated as knowledge sources.
  (iv) Conducting laboratory experiments is a method to generate knowledge power for organization.
   (v) Intellectual property under transfer of technology (TOT):
       Technical expertise obtained from TOT projects could be considered for implementation for other projects of organization.

The internal sources of organizational knowledge are complemented by external sources. The sources address changing trends in external environment and enable to update existing knowledge of organization. Some of the external sources are:

   (i) Military and other international standards for industrial processes.
  (ii) Application information published product manufacturers.
 (iii) Examining the products of competitors.
  (iv) Access to reputed digital libraries.
   (v) Attending or organizing the programs conducted by professionals.
  (vi) Attending conferences.

The knowledge obtained from internal and external should be documented and channelized to improve processes and products. Computerizing organizational knowledge is an effective method to channelize it for improvement.

4.5 Competence

Competence is the ability of an individual to perform work as per specified needs. Most organizations maintain individual records for the competence of personnel with details of date of joining, education, past assignments before joining the organization, changes in internal assignments with dates, current assignment in organization, experience acquired from the assignments, training programs attended by the personnel and other administrative information. The individual records of personnel are updated for subsequent changes in assignments and for participation in competence improvement programs.

The competence requirements of personnel that affect the performance and the effectiveness of QMS and the competence levels of existing personnel from their individual records are analyzed to determine the competence gap. If gap exists, actions are taken to acquire the necessary competence and the effectiveness of the actions taken is evaluated for closing the competence gap. Appropriate documented information is maintained for the competence gap, actions taken and evaluating the effectiveness of the actions taken. Suggested actions to acquire competence and evaluating the effectiveness of the actions taken are explained.

4.5.1 Acquiring Competence and Evaluating Effectiveness

Actions appropriate for personnel are planned and implemented to impart the identified competence needs. Competence needs are achieved if the actions taken are effective. The effectiveness of the actions taken is generally evaluated by measuring the work performance of the personnel. Simulated exercises or quiz programs could be used to complement the evaluation methods. The suggested actions to acquire competence are:

(i) Re-assignment of currently employed persons.
(ii) Training programs by internal or external experts.
(iii) On-the job training for operators.
(iv) Nominating persons for attending formal educational courses.
(v) Recruiting competent persons.

4.6 Awareness

After establishing QMS, process owners organize and conduct a formal awareness program for their personnel. The module for the program contains:

(i) Overview of ISO 9001 QMS.
(ii) Quality policy.

(iii) Quality objectives.
(iv) Contribution of personnel to the effectiveness of QMS, including the benefits of improved performance.
(v) The implications of not conforming with the QMS requirements.

Process owners conduct the awareness program for new entrants also. Organizations maintain documented information as evidences for conducting the program.

4.7 Communication

The requirements of communication are part of the documented information (quality procedures) of QMS. Responsibilities for processes and sequence and interaction of the processes described in quality procedures addresses the requirements of internal and external communications providing answers for the questions, what, when, with whom, how and who.

The communication needs of QMS such as ensuring customer focus, quality policy and objectives and defining roles, responsibilities and authorities are addressed in quality manual. They are generally communicated by top management and management personnel. Methods like notice boards and in-house news bulletins and information technology methods (e-mails and intranet) are also used for communication.

4.8 Documentation Requirements

QMS documents are prepared considering the requirements of ISO 9001 and organization for ensuring satisfactory operation and effectiveness of QMS processes. The QMS documents maintained by organization are:

 (i) Quality manual (Sect. 2.6.1).
 (ii) Documented information (quality procedures) for generic QMS processes.
(iii) Product related engineering documents.

QMS documents are created, updated and controlled by the relevant functions of organization. For example, product related engineering documents are created, updated and controlled by design team. Appropriate arrangements could be decided for other QMS documents. The considerations specified in Sect. 2.1.2 are applied for creating the documents.

4.8.1 Creating and Updating

The administrative and technical requirements for creating QMS documents are explained. The documents are updated as needed by corrective and improvement actions. The general procedure for creating documents is applied for updating the documents. The requirements for creating QMS documents are:

(i) Identification and description:

These are administrative requirements. The title of document, document number, date, version status and identification of the persons responsible for preparation, review and approval are indicated in QMS documents. When documents are updated, change control information is indicated in the QMS documents.

(ii) Format and media:

These are administrative requirements for creating and maintaining QMS documents. Format represents the language, figures, descriptive method and flow charts as appropriate for QMS documents. The media for the documents could be paper or software. Organizations could use software for preparing, maintaining and updating the documents. There is master control of the documents and rights are given to users using software to access their relevant documents. Distribution and updating of QMS documents is considerably simpler with software media.

(iii) Review and approval:

QMS documents are reviewed for suitability and adequacy. The review requirements are:

- Identification and description of document.
- Complying with ISO 9001 requirements.
- Addressing quality needs of internal and end customers (Sect. 1.3.1)
- Avoiding contradicting information.
- Guidance for proper use, as appropriate.
- Legibility of the documents.
- Updating related documents, as applicable.
- Appropriate information for external providers (purchasing), inspection, production, and preservation in product related engineering documents.
- Version status.

Process owners review and approve their relevant documents. The documents are released after review.

4.8.2   Control of Documented Information

Procedures for controlling the documented information of the QMS processes are
maintained. The control procedures are established to ensure:

(i) Availability and suitability of documents:
     The documents are available and suitable for use, where and when needed.
     The current revision of documents is made available to process owners to
     ensure the suitability of the documents.
(ii) Protecting documents:
     Product related documents might require protection from loss of integrity,
     improper use, etc. Appropriate arrangements such as issuing the documents
     when required and returning the documents after processing could be made by
     organization. Access to engineering documents is also controlled to ensure
     that the documents are available for authorized users.

4.8.2.1   Procedure

Organization maintains separate procedures for the control of documented information for the generic QMS processes and for the product related documents. The
procedures address the following requirements, as applicable:

(i) Distribution, access, retrieval and use:
     Information for distributing documented information to users is either part of
     procedures or defined separately. The users organize the documented
     information for easy access, retrieval and use.
(ii) Storage and preservation:
     Documented information is stored and preserved using appropriate means,
     including preservation of legibility.
(iii) Control of changes:
     Organization maintains a master list of documented information for generic
     QMS processes and product related processes for controlling the documents.
     Improvement and corrective actions might result in changes in the docu-
     mented information for the processes. The reason for changes is examined
     and the details of changes in documents are identified. The changes are
     reviewed and approved. The relevant documents are amended, incrementing
     the version status of the documents. The new version status of documents is
     reflected in the master list of documents. Work instructions for the control of
     changes could be maintained.

(iv) Retention and disposition:
Obsolete documented information is usually removed form users except one set of the documents are preserved for knowledge purposes. Retention period for retaining documented information to provide evidences for demonstrating compliance to QMS is decided and it could be indicated in the respective records or separately.

(v) Documented information of external origin:
Documented information of external origin is military, telecom and other international standards and they are used by design, manufacturing and QA functions. Referring on-line Standards store is a method to obtain the current status of documents of external origin.

4.8.2.2   Quality Records

Organization retains documented information (quality records) to demonstrate compliance to the quality procedure for creating, updating and controlling QMS documents. The records are protected from unintended alterations. Appropriate information technology or other suitable method could be for the protection. The quality records are:

(i) Approved QMS documents.
(ii) Distribution or access list for the documents.
(iii) Master list of internal documents.
(iv) Change control information for updated documents.
(v) Master list of external documents with verification of version status.
(vi) Retaining obsolete documents with suitable identification as specified in process procedure.

References

1. Role of Measurement and Calibration, UNIDO, Vienna, V.05-90773, 2006
2. FDA Regulation 21CFR820.72 Inspection, measuring, and test equipment, title 21, vol 8, Apr 2015
3. Mike cable, calibration: a technical guide. The International Society of Automation, USA, 2005
4. Measurement traceability and calibration in the mechanical testing of metallic materials. UK Accreditation Service, Publication # 24, edn 1, Nov 2000
5. Guidance on traceability, A61-02, Revision 1.2, Canadian Association for Laboratory Accreditation, Oct 2012
6. ISO 9000:2005, Quality management systems-fundamentals and vocabulary
7. Thabo Chesalokile, verification versus calibration, South African National Accreditation System, Report-7, 2011

8. Calibration laboratories, Technical guide for DC low frequency measurements. NIST Handbook150-2A, March 2004
9. Guidance for the validation of analytical methodology and calibration of equipment used for testing of illicit drugs in seized materials and biological specimens. UNDOC, ISBN 978-92-1-148243-0, Oct 2009
10. Calibrating the agilent 3458A with the 5730A, multifunction calibrator. Application Note 1264255E_EN, Fluke Calibration, USA, Sept 2013

Chapter 5
Operational Processes

Abstract Operational processes contribute significantly for the growth of organization. Planning, understanding customer requirements for products, design & development, control of external providers and production are the operational processes. The requirements of the operational processes and other related processes are explained with practical examples for deeper understanding. The other related processes are identification and traceability, customer property, preservation, post-delivery activities, control of changes, release of products and nonconformity control.

5.1 Operation

ISO 9001 specifies a set of generic operational processes for converting customer requirements into products. The generic operational processes interact with management, support and product specific processes to realize products. The generic operational processes of ISO 9001 are:

  (i)   Operational planning and control.
 (ii)   Requirements for products and services.
(iii)   Design and development of products and services.
 (iv)   Control of externally provided products.
  (v)   Production and service provision.
 (vi)   Release of products and services.
(vii)   Control of nonconforming outputs.

Organizations maintain documented information (quality procedures) for all the generic operational processes to ensure that the operations of the processes are effective for satisfying the requirements of products and customers. Adequate guidance and examples form industries are provided for preparing quality procedures for the generic operational processes.

© Springer International Publishing AG 2017

D. Natarajan, *ISO 9001 Quality Management Systems*,
Management and Industrial Engineering, DOI 10.1007/978-3-319-54383-3_5

Fig. 5.1 Operational
planning and control

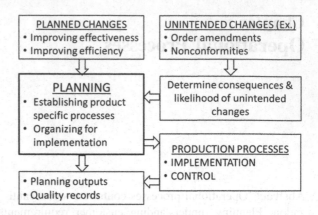

5.2   Operational Planning and Control

Product specific processes are required for the provision of the products. Examples
of product specific processes for RF Filters are shown in Fig. 1.1. Operational
planning and control require that product specific processes should be planned,
implemented and controlled for:

(i)   Meeting product conformity and delivery requirements for the provision of
      products.
(ii) Implementing the actions determined to address risks and opportunities for the
      processes.

   The requirements of planning, implementation, control and outputs are shown in
Fig. 5.1. Additional information is provided for the requirements. Addressing
planned and unintended changes is also presented.

5.2.1   Planning

Planning is necessary for establishing product specific processes. Establishing refers
to preparing work instruction documents for processes. Organizing for the implementation of the processes is also part of planning. Establishing the processes and
organizing for the provision of products are explained.

5.2.1.1   Establishing Product Specific Processes

The generic QMS processes of organization are established by preparing the
required documents (quality manual and quality procedures) for the processes
conforming to the general requirements specified in Sect. 3.5. The same general

requirements are applied for preparing product specific work instruction documents also. The work instructions are prepared by design team defining the controlled conditions specified in Sect. 5.7. Additional considerations for preparing work instruction documents using the ideas of Indian classic literature are presented in Chap. 9. The general requirements for preparing the documents are:

(i)   Inputs including the resources needed to achieve conformity to products.
(ii) Work instructions for the operations of process.
(iii) Acceptance criteria for in-process and finished products.
(iv) Controls for the process in accordance with the criteria.
(v)   Integrating the actions to address risks and opportunities with production process related procedures and documents. Integrating the actions with QMS processes is explained for three operational processes in Sects. 3.7–3.10.

5.2.1.2   Organizing for Implementation

Organizing for the implementation of production processes to provide products meeting customer requirements is the challenging activity of operational planning. It requires the co-ordination of many other processes. Planning for the implementation and control of production processes requires:

(i)   Interacting with marketing for the requirements for products (Sect. 5.3).
(ii) Interacting with design and development for new products (Sect. 5.4).
(iii) Verifying the status of in-house stock levels and work-in-progress with internal processes and external providers.
(iv) Organizing and controlling externally provided processes and products (Sect. 5.5).
(v)   Organizing and controlling internal production processes (Sect. 5.7).
(vi) Processing the property belonging to customers (Sect. 5.9) for production processes.

5.2.2   Implementation, Control and Outputs

Most organizations provide ERP software support for planning, implementing and controlling internal and external production processes. The processes are implemented, monitored and controlled by the process owner of operational planning in accordance with the product related engineering documents.

5.2.2.1   Outputs

The outputs of planning process are internal orders for production processes with inputs and resources, supported by appropriate product related engineering documents. Purchase orders for external providers with appropriate purchasing information are also the outputs of planning. If the commitment of customers could not be met, appropriate information is provided for communicating to customers.

5.2.2.2   Quality Records

Completed internal orders for production processes and purchase orders for external providers are quality records. Verification records for externally provided products are maintained as per Sect. 5.5.5.5. The quality records demonstrate that the processes have been carried out as per QMS and product related engineering documents.

5.2.3   Addressing Planned and Unintended Changes

Changes are intended to be beneficial to organization and they need to be carried out as determined by the organization, considering risks and opportunities caused by the changes [1]. Changes in operational planning could be categorized as planned changes and unintended i.e. unplanned changes. The categories of the changes and the actions to mitigate adverse effects caused by unintended changes should be understood for managing changes.

5.2.3.1   Planned Changes

Planned changes occur when organization recognizes the need for improving the effectiveness and efficiency of the operational planning and control. Examples of planned changes are improving the existing ERP software, organizing competence improvement programs for managing the quality objectives of operational planning and re-defining minimum stock levels for products.

5.2.3.2   Unintended Changes

Unintended changes are forced changes on operational planning caused by internal and external factors. Organization is expected to react and implement appropriate measures in planning (Sect. 5.2.1.2) to deal with the changes. Sources of unintended changes are nonconforming products in production processes, customer order amendments and performance failures of external providers. Unintended changes are reviewed and controlled for meeting customer requirements and product conformity.

All the unintended changes that are generated need not be considered for initiating corrective measures in planning. Some changes could be ignored. The triggered changes are reviewed by determining the consequences of the changes and the likelihood of the consequences [1]. The classification of risk levels and the probability of risk occurrence explained in Sect. 3.7.1 could be applied for considering or ignoring unintended changes. Consequence of unintended change represents the impact on meeting customer requirements and product conformity, if appropriate measures are not implemented for the changes. Likelihood of the consequence is the probability of failure to achieve the goals if appropriate measures are not implemented for the changes. Unintended changes having minor impact on the goals with low probability of occurrence could be ignored.

5.3 Requirements for Products and Services

The requirements of products are determined using customer specified requirements. The determined requirements are reviewed and finalized. The finalized requirements of products are the inputs for design and development of the products and operational planning. The general procedure for finalizing the requirements of products is shown in Fig. 5.2 and the requirements of the procedure are:

(i)   Customer communication.
(ii) Determining the requirements for products.
(iii) Reviewing the requirements for products.
(iv) Addressing changes to the requirements for products.

Fig. 5.2 Finalizing requirements for products

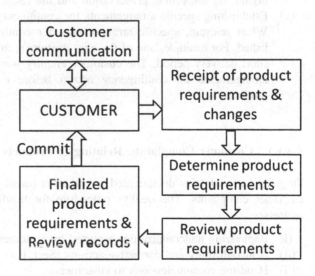

5.3.1 *Customer Communication*

Communications with customers include:

(i) Providing information relating to products:
 Customers require product related information for various applications and
 organization should provide the information. Some of the arrangements to
 provide information are product catalogues and application notes for products. Free samples could be provided for the developmental projects of
 customers.

(ii) Handling enquiries, contracts or orders including changes:
 The requirements for products are communicated by customers to organizations in many ways. Enquiries might be verbal or in written form.
 Customers might release repeat orders directly without formal enquiries
 based on previous quotations. Organization processes enquiries, contracts or
 orders and changes from customers and sends quotations and order acceptance to customers accordingly.

(iii) Customer feedback and complaints, relating to products:
 Organization maintains documented information for obtaining customer
 feedback relating to products. Obtaining customer feedback is part of the
 quality procedure for customer satisfaction (Sect. 6.3). The procedure for
 handling customer complaints is explained in Sect. 5.3.1.1.

(iv) Handling or controlling customer property:
 Customer property, when received, is identified. The information and the
 customer property are sent to the process owner of operational planning for
 organizing inspection, preservation and use (Sect. 5.7).

(v) Establishing specific arrangements for contingency actions:
 When relevant, specific arrangements for contingency actions are established. For example, one of the elite customers might require products with
 short delivery period. The customer enquiry warrants establishing specific
 arrangements for contingency action before committing to supply the
 products.

5.3.1.1 Customer Complaints, Relating to Products

Organization maintains documented information (quality procedure) for handling
customer complaints. The quality procedure for handling customer complaints
addresses:

(i) Registering nonconformities reported by customers.
(ii) Nonconformity and corrective actions (Sect. 6.8).
(iii) Handling communications to customers.
(iv) Closing customer complaint.

Software support is essential for managing customer complaint data efficiently. The administrative and technical requirements of failure data management system for electronic product are illustrated with examples in reference [2].

5.3.2 Determining the Requirements for Products

The requirements for the catalogue products of organization are completely defined and no further efforts are needed to determine the requirements of the products. The products could be offered directly to customers as per enquires and orders. Electronic components and commercially off-the-shelf (COTS) items are examples of catalogue products. The requirements of new products should be determined.

5.3.2.1 Statutory Requirements of Products

Statutory requirements are the laws of governments. For example, many governments have promulgated laws prohibiting the use of heavy metals like lead and mercury in products and the legal provision is known as RoHS (Restriction of Hazardous Substances). If statutory requirements are applicable for products, they are documented. Statutory requirements are applicable for electronic equipment and might be applicable for components.

5.3.2.2 Regulatory Requirements of Products

Governments create agencies with powers to regulate products, protecting the interests of customers. The regulatory requirements could be demanded by customers or they could be specified by organizations. Controlling the Radio Frequency (RF) emission levels of mobile phones is an example of regulatory requirement. The supply of mobile phones is regulated to comply with the essential requirements of the directive, 1999/5/EC for the avoidance of interference or matters relating to public health. Regulatory requirements are usually applicable for electronic equipment and systems.

5.3.2.3 Product Requirements Identified by Organization

Organization identifies requirements that are necessary to ensure reliability and producibility of product. Examples of the requirements are additional environmental tests and standardization of parts. The identified requirements are usually documented as design and development inputs (Sect. 5.4.2). Organization documents what could be met and what could not be met for the products it offers when examining the product requirements of customer.

5.3.3   Review of Product Requirements

Organization reviews the determined requirements related to products before sending quotes to customer enquiries or submission of tenders. If customer orders are received directly without enquiries, the reviews are performed before sending order acceptance. When customer does not provide a documentary statement of their requirements i.e. when organization receives verbal enquiries, the customer's requirements are confirmed and documented by the organization.

Formal review is conducted for the requirements of new products before committing to supply the products. The process owners of design, operational planning and other relevant processes review product requirements before committing to supply products to customers. External experts might be invited for reviewing complex projects. The review team focusses on feasibility, completeness and identifying efficient methods of complying with product requirements. The results of review are documented.

The requirements of catalogue products (Ex.: Electronic components and COTS items) are completely defined. Hence, their review could be limited to the delivery requirements of customers.

5.3.3.1   Ability of Organization

The ability of organization to meet the requirements for products to be offered to customers is ensured. The availability of technology, resources and competent personnel are examined for meeting the requirements of products. Organizing external resources for meeting the requirements is acceptable. The requirement is generally applicable for new developmental projects that are outside the product range of organization. For example, organization having the product range of microwave cavity filters should assess their abilities before quoting for stripline or microstrip RF Filters.

5.3.3.2   Requirements Specified by Customer

Product requirements specified by customers, including the requirements for delivery and post-delivery activities are reviewed before commitment. Requirements for inspection and installation, training and providing spares with product are some of the delivery and post-delivery activities. Customers might indicate the supply of unique items (customer property) for use in the manufacturing of products.

5.3.3.3   Legal and Other Requirements, Identified by Organization

The requirements not stated by customer but necessary for the specified intended uses are determined as per Sect. 5.3.2.3. The requirements and the applicable statutory and regulatory requirements are reviewed.

5.3.3.4   Resolving the Differences

Quotations or submission of tenders are sent by organizations to customers after reviewing the requirements of products. Customers examine them and convert them into orders. When customer orders are received, organizations review the orders to identify the requirements of products, differing from those submitted in quotations or submission of tenders. Action is taken by organizations to resolve the differences before accepting orders or contracts.

5.3.3.5   Retaining Documented Information

Documentary information (records) is retained to provide evidences for conducting reviews. The evidences that are retained are:

(i)   Resolving differences (Sect. 5.3.3.4):
      Records are not necessary, if there are no differences to resolve. As organizations use ERP software for processing customer requirements, it is recommended to maintain suitable records even if there are no differences to prevent human errors (Sect. 5.7.5).
(ii) Reviews for new products:
      The results of review are maintained in appropriate form. For larger products, project proposal could be maintained.
(iii) Distribution of review output documents:
      The review output documents are distributed to relevant persons for implementation. Generally, the relevant persons would be the process owners of design and development and operational planning and control.

5.3.4   Changes to Requirements for Products

Changes to requirements for products might be initiated by customers or organization. Changes are also reviewed before commitment. The finalized changes are indicated suitably by amending the review output documents. The amended documents are also distributed to relevant persons.

5.4 Design and Development of Products

Design and development practically decide the total cost of products and the level
of customer satisfaction. Design and development processes are initiated by
the commitment of organizations for delivering products to customers or by the
directive of top management. Design refers to the processes from the receipt of the
requirements of products to the preparation of provisional engineering documents.
Development refers to fabricating proto-models of products using the provisional
documents and evaluating the proto-models for conformance against product
specifications. The outputs of development are the set product related engineering
documents for internal and external customers. Design and development processes
are established, implemented and maintained to ensure subsequent provisioning of
products. Organization maintains documented information (quality procedure) for
the processes and adequate information is provided for establishing the quality
procedure.

5.4.1 Design and Development Planning

Design and development planning refers to identifying the stages and controls for
the stages to design and develop products. Stages that are appropriate for product
development are decided by design team. Controls are reviews, verification and
validation activities to ensure that the planned requirements of the stages are met.
Organization usually maintains documented information (quality records) for the
output of planning, monitoring the stages and controlling the outputs of the stages.
The considerations for determining the stages and controls for the design and
development of products are explained. Additional considerations for design and
development planning using the ideas from Indian classic literature are presented in
Chap. 9.

5.4.1.1 Nature, Duration and Complexity

The nature of activities, project duration for completing the activities and the
complexity of the activities for the design and development of products are considered during planning. Nature of activities represents the type of design and
development activities. For example, mostly internal activities are associated with
the design and development of RF Filters whereas internal and external activities
need to be coordinated for developing electronic equipment. Complexity represents
technology and efforts associated with the activities. Design and development plan
for electronic systems would be more elaborate compared to that for RF Filters.

5.4.1.2 Determining Process Stages with Reviews, Verification and Validation

Many micro-level activities are involved during the design and development of products. The activities could be searching digital libraries for understanding trends in design, examining the capabilities of competitors, circuit design, selecting parts, etc. They are relevant activities and they are performed. However, all the activities are not listed as the stages of planning. Only the macro-level stages of design and development are identified and documented for meaningful planning.

Review, verification and validation are controls and they are interfaced with the design and development stages appropriately. In general, the activities ensure that the relevant stages are completed with the confidence of satisfying the requirements of products and customers. Additional information regarding the controls is provided in Sect. 5.4.3.

A typical plan for the design and development of RF Filters indicating the stages interfaced with review, verification and validation activities is shown in Fig. 5.3. The stages with controls are:

(i)   Preparing design and development inputs.

  – Review

(ii) Product design.

  – Review

(iii) Preparation of provisional engineering documents.
(iv) Manufacturing proto-models.
(v)   Preparing verification and validation test plans.

  – Review

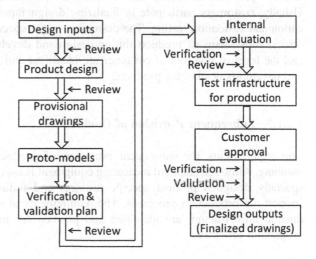

Fig. 5.3 Typical design and development plan for RF Filter

   (vi) Internal evaluation.

         – Verification
         – Review

  (vii) Identifying test infrastructure requirements for production
 (viii) Customer approval

         – Verification
         – Validation
         – Review

   (ix) Updating and releasing of finalized engineering documents.

5.4.1.3 Responsibilities, Authorities, Resource Needs and Interfaces

A team of engineers is involved to implement the planned process activities for the
design and development of products. Project manager assigns the activities to
individual engineers or to a group of engineers. The manager defines responsibilities and administrative authorities with time schedules for completing the assignments. The interfaces between the personnel of the design team are considered
when defining the responsibilities and authorities. The internal and external
resources needed for performing the activities are identified.

5.4.1.4 Customer Involvement and Controls

Customers or their authorized representatives and other interested parties specify
their involvement and controls for the design and development of aerospace and
military products or safety critical products for chemical and nuclear plants.
Usually, customers participate in finalizing design inputs, design reviews, verification and validation testing. The customers might specify controls for inspecting
critical subsystems of products during design and development. The involvement
and the level of controls of customer are identified and integrated into the design
and development plan for products.

5.4.1.5 Subsequent Provision of Products

The requirements for subsequent provision of products are considered during
planning. Although standard measuring equipment is used for production processes,
specially designed product specific test jigs and fixtures might be required to
support the production processes. The requirements of such product specific production infrastructure are identified and planned for making them available for
production processes.

5.4.1.6   Documented Information for Demonstration

Proto-models of products are manufactured during the development phase of the products. Documented information (test methods) is prepared for evaluating the proto-models to demonstrate that the design and development requirements have been met. Evaluation program for proto-models consists of internal verification tests and customer validation trials. As the complexity of product increases, elaborate evaluation program would be needed involving internal and external facilities. Hence, design and development plan addresses product evaluation program to complete projects as per schedule. Preparing documented information in the form of detailed verification and validation test methods and test record formats is part of design and development plan.

Practical difficulties exist to prepare product evaluation documents at the beginning of design phase. The activity for preparing the documents is shown in the design and development plan indicating its status appropriately. The documents are prepared at appropriate stage of product design and the plan is updated as the design progresses.

5.4.2   Design and Development Inputs

Design and development inputs guide the processes from planning to the release of finalized engineering documents for production. The essential requirements, specific for the design and development of products are determined and documented as design inputs. The considerations for determining the requirements for the inputs are explained.

5.4.2.1   Functional, Performance, Statutory and Regulatory Requirements

The functional, performance, statutory and regulatory requirements of products, finalized after reviewing the requirements specified by customers are the basic design and development inputs (Sect. 5.3). Similar requirements are defined for developing products, initiated by top management.

5.4.2.2   Information Derived from Previous Similar Designs

Organization preserves design files of products for knowledge purposes. Learnings from previous designs of similar products provide useful information for new designs. Proven design ideas, using standard modules and verification test plans of similar products could be considered for new designs. The information derived from similar designs is useful for reducing design cycle time and improving producibility. Such information is extracted from previous designs and documented as design inputs.

Organizations conduct root cause analysis of failures observed during manufacturing and those reported by customers in field applications. The causes and the corrective actions to eliminate the recurrence of the failures in previous similar designs are information feed-forward for new designs and the information is suitably documented as design inputs.

5.4.2.3 Additional Considerations and Documentary Evidences

International standards and codes specifying design practices exist for products. Military standards exist for the design and development of defense equipment. Applicable standards or codes are documented as design inputs.

The potential failures of products and their consequences are documented as design inputs. A team of engineers is involved for the design and development of product. Project manager conducts brain storming session synergizing the expertise of engineers to identify possible (potential) failures of product during manufacturing and usage. Potential failures that could be caused by users due to the inadvertent misuse of products are also considered in the analysis. The consequences of the failures are also identified.

Design team ensures that the inputs for the design and development of products are adequate, complete, unambiguous and not in conflict with each other. The design and development inputs, the considerations used for determining the inputs and the basis of the considerations are documented.

5.4.3 Design and Development Controls

Design and development of product is one of the critical processes of organization. Controls are needed to ensure that the planned results of the process are achieved. Design and development controls are review, verification and validation. Design team decides the appropriate controls for the stages of design and development, considering the complexity of products. The controls are interfaced with the activities of design and development plan as shown in Sect. 5.4.1.2. Although design review is repeatedly interfaced with many stages of the plan, the objectives for the design reviews are different. Hence, the results to be achieved by the design and development control activities should be defined.

5.4.3.1 Design Review

Design review is retrospection of a design and development stage after completing the activities of the stage. The objectives of each design review are derived from the requirements of the design and development stage for which the design review is performed. Also, the composition design review team depends on the objectives of the review. Two examples are provided for understanding design review, using the design and development plan of RF Filters in Sect. 5.4.1.2.

Design review is conducted after determining the design and development inputs for RF Filters. The primary objectives of the review are to obtain information for design approach, observed and potential failures in the production of similar products and customer feedback. Personnel of design and manufacturing functions participate in the review. Project team presents the documented design and development inputs in the review meeting to stimulate the generation of new inputs from the participants.

Organization conducts internal verification tests on the proto-models of product before submitting the products to customers for obtaining approval. Design review is conducted after internal verification testing also. The review focuses on the adequacy of the verification tests and the results obtained to ensure first time approval by customers. Design teams and quality assurance personnel who have knowledge on standards and codes participate in the review. Documentary information is retained as evidences for the results (decisions) of reviews and for implementing the decisions of the reviews.

5.4.3.2 Design Verification

Design verification is performed to confirm that design and development outputs satisfy the design and development inputs of products. Simulation software packages are available for products (Ex. Stripline RF Filters) and they could be used for the design and verification of the products before preparing provisional engineering documents. Design verification is also performed on the proto-models of products. Dimensional, electrical and environmental tests are examples of verification tests conducted on RF Filters to ensure first-time approval by customers. The tests demonstrate compliance to the design and development inputs of products. Necessary actions are taken on the problems observed during verification tests. If required, relevant verification tests are performed again for confirming compliance. Documentary information is retained as evidences for the results (findings) of verification tests and for the actions taken. Design verification tests generally precede design and development validation tests.

5.4.3.3 Design Validation

Design validation is performed to confirm that the resulting product is capable of meeting the specified application or intended use. Intended use is the actual operating environment for product. Design validation is usually performed by customers on the proto-models of products. For example, customers mount the proto-models of RF Filters in relevant equipment and verify the performance of the equipment under various operating environments in validation testing. Satisfactory functional performance in various use operating conditions demonstrates the acceptance of products in validation testing. Customers might repeat performance verification tests also before conducting validation tests.

Customers might involve organization in performing validation tests, considering the complexity of products. Necessary actions are taken on the problems observed during validation testing. If required, relevant validation tests are performed again for confirming compliance. Documentary information is retained as evidences for the results (findings) of validation tests and for the actions taken.

5.4.4 Design and Development Outputs

The outputs of design and development of products are the proto-models and the finalized engineering documents of the product. The engineering documents are prepared after confirming that the proto-models of product meet the requirements of design and development inputs as evidenced by the results of verification and validation tests. Finalized engineering documents are released for users as per the procedure in Sect. 4.8. The engineering documents are the evidences for the outputs of design and development of product. The documents are released after ensuring:

(i) Adequacy of information for subsequent processes:

– Information for external providers (Sect. 5.6)
– Defining product specific controlled conditions:
    Controlled conditions for production and service provisioning are defined in Sect. 5.7. The controlled conditions that are appropriate and specific to products are described adequately in product related engineering documents.
– Monitoring and measuring requirements and acceptance criteria:
    They are defined in product related engineering documents and they are part of controlled conditions.

(ii) Characteristics of products for intended purpose:
    Appropriate documents specify the characteristics of products that are essential for use, safe operation. Relevant information for maintenance is also specified.

Organizations design and develop the methods of providing services to customers. Design output documents provide information for delivering services to customers. Automobile service stations and banks are examples of providing services to customers.

5.4.5 Design and Development Changes

The need for changes arises during or subsequent to the design and development of products. Provisional engineering documents are updated for the changes observed during design and development. Released engineering documents are updated for the changes reported by internal and external customers. Documented information (quality records) is retained to provide evidences for:

(i) Record the details of proposed changes:
   The source of proposed changes, reason for changes, the proposed changes and the document reference are recorded. The changes could arise from the results of internal verification testing or production processes or customer reported failures.

(ii) Evaluating the changes:
   The proposed changes are reviewed to ensure that there is no adverse impact on product conformity. If required, appropriate verification test are conducted before confirming the need for the changes. The results of reviews and verification tests are maintained.

(iii) Implement the changes with proper authorization:
   The authorized person of design team amends the engineering document, maintaining the history of changes.

5.5 Control of Externally Provided Processes and Products

Every organization is dependent on external providers for meeting the requirements of customers. Externally provided item could be processes, products or services. Externally provided process is popularly known out-sourced process, where organization supplies all the input materials for converting them into specified product. Supplying all electronic components for converting them into assembled PCBs is an example of externally provided process. Externally provided products are bought-out items. The items could be components like ICs, raw materials or COTS items. Products include services also. Obtaining PCB layout design as per schematic information and utilizing the services of test centers for product evaluation are examples of externally provided services.

5.5.1 Applicability of Controls

Organization should control externally provided processes and products including services for ensuring conformance to the requirements of the organization. The controls on the products and services of external providers for utility purposes (Ex. Photo-copiers) are not part of the requirements of QMS. Controls on the processes, products and services of external providers are applicable when:

(i) Products and services (Ex. Raw materials and parts) from external providers are intended for incorporation into the organization's own products and services.

(ii) Products and services are provided directly to customers by external providers on behalf of organization. Organization might instruct external providers to deliver spares directly to customers.

(iii) A process or part of process (Ex. Semi-finished products) is provided by an external provider as a result of a decision by organization.

5.5.1.1 Pre-requisites for Controls

The controls to be applied on external providers are determined primarily from the requirements of products and customers. Before determining the type and extent of control, potential external providers are evaluated and selected based on their ability to provide processes, products and services conforming to the requirements of organization. Approved external providers are monitored for performance and re-evaluated at appropriate intervals. The activities and the criteria for the activities are explained. Documented information is retained as evidences for the implementation of the activities.

5.5.2 Evaluation of External Providers

The general procedure for evaluating external providers is shown in Fig. 5.4. Criteria to assess the abilities are identified as appropriate for external providers. The general criteria for the assessment are:

(i) Delivering required quantity of items with consistent quality.
(ii) Lead time for delivery.
(iii) Adherence to committed delivery schedule.
(iv) Providing support after the delivery of items.
(v) Additional product related criteria required by organization.
(vi) Commercial requirements.

Fig. 5.4 Evaluation of
external providers

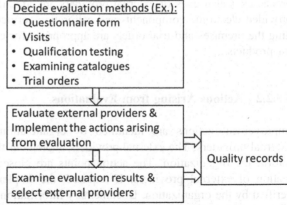

5.5.2.1   Evaluation Methods

Evaluation methods are formulated reflecting the criteria elements for assessing the capability of service providers. The suggested evaluation methods are shown in Fig. 5.4 and the methods are:

(i)   Sending questionnaire form:
      Questionnaire form is sent to external providers preferably prior to visiting their premises. The form is designed to obtain information regarding people, infrastructure and information on quality management system for operations.

(ii) Visiting the premises of external providers:
      Visiting the premises of external providers is used to confirm the information obtained through the questionnaire form and to identify the areas for improvement.

(iii) Qualification testing of items:
      Items of external providers are subjected qualification tests as per relevant standards for verifying the capabilities of external providers.

(iv) Catalogues of external providers:
      The catalogues of external providers indicate the capabilities of their organization and items. The information provided in the catalogues is used for evaluating external providers for simple components.

(v)   Trial orders:
      Organization could release trial orders for products on external providers with appropriate controls and conditions for assessment.

Appropriate method or a combination of the methods is decided by organization for evaluating external providers. For example, questionnaire form and visiting the premises of external providers are appropriate for evaluating externally provided processes. Catalogues of external providers are adequate for evaluating externally provided electronic components and calibration services. Questionnaire form, visiting the premises and trial orders are appropriate for evaluating external providers for products.

5.5.2.2 Actions Arising from Evaluations

Improvement actions are identified by organization during the evaluation of external providers. The external providers implement the improvement actions and inform the organization. The action points are closed after reviewing the information of external providers. If required, the implementation of the actions is verified by the organization. Documentary information is retained as evidences for the actions arising from the evaluations and their implementation.

5.5.3 Selection and Monitoring

Selection of external providers (suppliers) is based on the results of evaluation and the suppliers are approved for providing processes, products and services, as applicable. Where feasible, approvals are accorded with broader scope of providing items. The provisions of qualified manufacturers for approving wide range of electronic components in military standards could be followed. For example, external providers for PCB assembly processes could be approved for wide range of assembled PCBs. Such approvals simplify purchasing process without compromising quality. The list of selected external providers is maintained and it is popularly known as Approved Suppliers List. Requirements for monitoring performance and other controls on external providers are also documented in the Approved Suppliers List.

5.5.4 Re-evaluation of External Providers

Approved suppliers list is maintained in spread sheet format indicating the name of supplier, the category of approval, date of approval, linkage to evaluation report, monitoring needs, etc. The validity of approval is also indicated and it is usually fixed as three years. Before the expiry of approval status, re-evaluation process is initiated. The re-evaluation procedure is same as that for the evaluation of supplier. The validity period and the rigor of re-evaluation procedure are decided by organization based on the performance of the supplier in the three year period.

5.5.5   Type and Extent of Control

Organizations decide the type and extent of controls on external providers to ensure that externally provided processes, products and services do not affect the organization's ability to consistently deliver conforming products to its customers. Examples of controls over suppliers are audits, review of historical data, monitoring, trending and inspection testing [3]. Documented process, change control, performing root cause analysis promptly to resolve issues related to products, PFMEA (Process Failure Mode and Effect Analysis), Statistical process control on critical variables and quality agreements are controls on suppliers [4]. The extent of the control is the level of the controls applied on the external providers. The controls are defined for externally provided processes, external providers and for the resulting outputs of externally providers.

5.5.5.1   Controls on Externally Provided Processes

Organization ensures that externally provided processes remain within the control of its QMS. For example, consider the outsourced PCB assembly process by original equipment manufacturer (OEM). The controls that could be applied by the OEM are:

(i)   Documented information defining the controlled conditions for the processes.
(ii)   Process change control.
(iii)   Quality agreement.
(iv)   Monitoring.

5.5.5.2   Controls on External Providers

Organization decides the controls that should be applied on external providers. Examples of the controls that could be applied on external providers are:

(i)   Approving multiple external providers for items:
      This is an indirect form of control on external providers, safeguarding the interests of organization.
(ii)   Review of historical data:
      Past performance of external providers is analyzed to derive useful control information.
(iii)   Conducting audits on the quality management system of external providers.

5.5.5.3 Controls on the Outputs of External Providers

Controls are necessary on the resulting outputs of external providers, delivered to organization. The controls that could be applied by organization on the items delivered by external providers are:

  (i) Providing test and measurements report with delivered items.
 (ii) Specifying that the concurrence of organization should be obtained for accepting nonconforming products with or without rework.
(iii) Specifying the application of statistical controls for the critical parameters of products.
 (iv) Verification of items at the premises of external providers.
  (v) Verification by organization after receipt of items.
 (vi) Specifying retention period for test records:
       UTAS A 9000 [5] specifies that the final acceptance data of aerospace products should be maintained for ten years.

5.5.5.4 Extent of Control

Organization is committed to meet customer and applicable statutory and regulatory requirements consistently. Externally provided processes, products and services could impact the organization's ability to meet its requirements. The extent of controls on the externally provided items is decided considering their potential impact on the ability of organization for meeting its requirements. For example, the control to obtain the concurrence of organization for accepting nonconforming products with or without rework is specified for high reliability applications such as aerospace products and it is not applied for all products. Verification of products is usually applied for all products.

External providers might have effective systems and controls for delivering quality items consistently and meeting applicable statutory and regulatory requirements. Organization could decide the extent of control on the external providers accordingly.

5.5.5.5 Verification or Other Activities on Externally Provided Items

Organization determines the necessary verification activities to ensure the externally provided processes, products and services to meet their requirements. Verification could be performed on 100% basis or on sampling basis. For example, 100% microscopic visual examination is performed for products used in critical applications.

Many OEMs eliminate or minimize verification activity and perform other types of control activities to ensure the externally provided items meet their requirements. The intent of controls is to provide quality products beyond what can be achieved through inspection and testing [3]. Co-operation between organization and external providers is necessary for eliminating or minimizing verification activity. Variations in the performance of items provided by external providers could be monitored for minimizing verification efforts.

Statistical methods are useful for monitoring variations in the performance of products. DPMO (Defects per million opportunities) is a statistical measure of expressing the number of defects (discrete data) observed in products and it could also be expressed as statistical sigma level. The mean and standard deviation of measured values (continuous data) of the critical characteristics of products could be computed and expressed as statistical sigma levels relative to their acceptance criteria. Standard calculators are available for computing DPMO and sigma levels. Sigma levels of four or more are acceptable for most products and the verifications of externally provided items could be eliminated. Four sigma levels can only be achieved by additional efforts in the design of products and processes.

5.6 Information for External Providers

The requirements of organization are communicated to external providers for designing, developing and manufacturing products as applicable. The requirements are also used by the organization for verifying the items provided by the external providers. Hence, the information for external providers is documented and controlled. Organizations decide the relevant information of products and ensure the adequacy of the information in the documents and in purchase orders prior to their communication to external providers. The purchasing requirements for procuring items from external providers are:

   (i) The processes, products and services to be provided.
  (ii) Approval of products and services.
 (iii) Approval of methods, processes and equipment.
  (iv) Approval of the release of products and services.
   (v) Competence, including any required qualification of persons.
  (vi) The external providers' interactions with the organization.
 (vii) Controls and monitoring of the external providers' performance to be applied by the organization.
(viii) Verification or validation activities that the organization, or its customer, intends to perform at external providers' premises.

5.6.1 Practical Considerations

Requirements of process or product or service are the basic information for delivering them by external providers. Engineering documents of organization define the requirements and they are made available to external providers. The documents are controlled by design team. If products need to be designed by external providers, engineering documents might not exist defining the requirements of products. Relevant extracts from contractual documents between organization and their customers are documented by the design team of the organization and the documents are made available to external providers for the design and development of items. Such documents are also controlled by the design team.

5.6.1.1 Additional Purchasing Information

Approvals, qualification of persons, interactions, controls, monitoring and verification are additional requirements that could be communicated to external providers. The additional requirements definitely ensure that externally provided items meet the requirements of organization consistently but they increase the cost of items also. Hence, they are selectively communicated to external providers for products used in high reliability and safety critical applications. The product specifications of customers are the basis for identifying the additional requirements of the products for communication to external providers by organization. Documented information is retained to provide evidences for communicating the requirements to external providers.

5.6.2 Requirements of Processes, Products and Services

The description of item and the relevant engineering documents completely specify the requirements of the item to be delivered by external providers. Copies of the documents are made available to external providers. Quantity, delivery date and other commercial requirements are communicated with the requirements.

5.6.3 Approval of Products and Services

Organizations specify the requirements of approval to national or international standards for the products to be delivered by external providers. For example, the approval requirement of UL (Underwriters Laboratory) or military standards is communicated to external providers for the supply of electronic components. Some of the tenders of government departments specify approval of products to their national standards as a condition for external providers to participate in tenders. Approval requirements quoting relevant standards are specified for the delivery of services by external providers.

5.6.4 Approval of Methods, Processes and Equipment

The requirements for the approval of methods, processes and equipment are generally applicable for high reliability and safety critical products. Examples are products for aerospace applications and safety critical chemical plants. Organizations include one or more of the requirements of approval as appropriate for the delivery of products by external providers.

5.6.4.1 Approval of Methods

Approved methods for monitoring processes, inspection, testing and other requirements could be communicated to external providers. Specifying standardized environmental test methods as per MIL-STD-810 [6] for the purchase of electronic sub-systems is an example for communicating approved methods.

5.6.4.2 Approval of Processes and Equipment

International Standards exist for most of the processes applied in the manufacture of products for high reliability and safety critical applications. The associated equipment for the processes is also controlled by the Standards. Referencing the Standard, ASME B31.3 [7] for welding and brazing for the manufacture of pressure piping used in petroleum refineries and chemical plants is an example of communicating the requirement for the approval of processes and equipment. Additional information is provided for understanding the approval of processes and equipment in Sect. 5.7.4.

5.6.5 Approval of the Release of Products and Services

Requirements for the approval of the release of products and services could be communicated to external providers. Examples of the requirements are:

(i) Submission of compliance reports along with products.
(ii) Informing nonconformities and corrective actions before delivering products.
(iii) Inspection and testing of products by organization at external provider's premises before delivery.

5.6.6   Competence and Qualification of Personnel

Competence is the ability of persons to perform work as per specified requirements. Organization could communicate competence needs of personnel for performing processes. For example, communicating that only trained operators should perform semi-rigid RF coaxial cable assembly operations is an example of competence needs of personnel. The requirement of qualification of personnel is communicated by organization when higher level of competency is needed for the critical processes of products used in high reliability applications.

Qualification of personnel means that only certified operators are allowed to perform an operation. It is training followed by certification in recognized institutions. IPC J-STD-001 [8] prescribes soldering practices and requirements for the manufacture of electrical and electronic assemblies and the standard is globally recognized. Institutions are available for qualifying operators in the methods and verification criteria for soldered interconnections as per the Standard. Communicating that only operators qualified to IPC J-STD-001 should perform soldering and verification operations in the manufacture of electronic assemblies is an example requiring the qualification of personnel.

5.6.7   Interactions, Controls and Monitoring

The objectives of interactions with external providers are to realize products and services as per the requirements of organization through co-operative efforts, which ensures success to both. The successful ones are more like collaborative partnership in which suppliers help their customers achieve their business objectives [9]. Interactions are similar to the reviews conducted at various stages during the design and development of products except that the interactions are planned to ensure collaborative partnership. The stages of interactions, applicable controls (Sect. 5.5.5) and monitoring performance are communicated to external providers.

Appropriate documented information is maintained by organization to provide evidences for implementing the requirements. Records are also maintained for the corrective actions for the nonconformities observed during the implementation of the requirements.

5.6.8   Verification and Validation

Verification is performing inspection and tests as per the engineering documents of organization. Validation is checking the functioning of products under use conditions. The requirements of verification and validation activities that the organization or its customer intends to perform at the external providers' premises are communicated. Verification or validation test plan is made available to external providers in advance. Documented information is retained to provide evidences for the verification and validation activities. Actions taken on the nonconformities observed during verification and validation are explained in Sect. 6.8.

5.7   Control of Production and Service Provision

Production is a set of product specific processes that are implemented for realizing products. Provisioning services (Ex. Services for automobiles) to customers also requires the implementation of set of service specific processes. Quality procedures are usually maintained for the control of production and service provisioning. Production and service provisioning are implemented under controlled conditions. The controlled conditions for manufacturing products are explained and the same could be applied for service provisioning.

5.7.1   Availability of Documented Information

Most of the controlled conditions for manufacturing products are specified in product related engineering documents. Some of the controlled conditions that are common to all products are specified in the generic quality procedures for production. The outputs of operational planning (Sect. 5.2.3) are the inputs for production processes. The outputs include product related engineering documents and supporting resources for production processes. The documents define:

(i)   Product characteristics:
      The characteristics of the product to be produced are defined in the product related engineering documents applicable for the production process.
(ii) The results to be achieved:
      The documents describe the results (acceptance criteria) to be achieved at the end of the process. If relevant, in-process acceptance criteria are also described in the documents.

5.7.2   Monitoring and Measuring Resources and Activities

Product related engineering documents indicate monitoring and measuring resources for implementing production processes. The resources are made available and are used for performing the activities of processes. The resources might be standard test equipment or gauges or in-house fabricated fixtures or a combination of them.

5.7.2.1   Monitoring and Measuring Activities

Monitoring and measurement activities are performed at appropriate stages as per product specific engineering documents. The documents specify the criteria for the control of processes and accepting the outputs of the processes. The results of the monitoring and measurement activities are recorded and the records provide evidences for:

(i)   The criteria for the controls are met.
(ii) The criteria for the outputs of processes are met.

5.7.3   Infrastructure, Environment and Competence of Personnel

Suitable infrastructure and operating environment are maintained for performing the activities of production processes. If special infrastructure and environment such as microscopes, fume chamber and radio frequency interference free enclosures are required, they are mentioned in product related engineering documents. The specified infrastructure and environment are utilized for performing the activities of processes.

5.7.3.1   Competence and Qualification of Personnel

Competence and qualification of personnel requirements, explained for external providers in Sect. 5.6.6, is applicable for internal production processes also. Appropriate competent and qualified persons are made available for performing the activities of the processes. The requirements of competent and qualified persons are specified in product related engineering documents.

5.7.4   Validation and Revalidation of Processes

Validation of process is defined as the qualification i.e. approval of the process. Production process needs validation if the resulting output of the process cannot be verified by subsequent monitoring and measurement to ensure the achievement of planned results. Such production processes are carried out using specific equipment, procedure and qualified persons and the requirements are decided during the qualification of the processes.

The requirements of validation and revalidation of processes are documented in the product related engineering documents. Understanding validation, identifying processes for validation, procedure for validation, international standards for validation and revalidation of processes are explained.

5.7.4.1 Understanding Validation of Processes

Non-destructive inspection and tests are conducted for verifying the characteristics of the outputs of production processes against criteria. For example, the output of PCB Assembly process is assembled PCBs. The characteristics (requirements) of assembled PCBs are solder connections with evidence of solder wetting and adherence and having smooth appearance [8]. Visual examination is conducted for the primary characteristics including for short circuit between adjacent tracks and damages to PCB or parts.

Tests are not feasible for the primary characteristics of the outputs of certain mechanical and metallurgical related production processes. For example, consider the outputs of arc welding process for aluminum alloys. The primary characteristic of the welded joint is its breaking load and it cannot be tested non-destructively for all the resulting products from the process. The welding process should be validated; otherwise, the defects of welded parts would be exposed when the parts are in use. The general procedure for validating processes is shown in Fig. 5.5.

5.7.4.2 Identifying Processes for Validation

Processes for validation should be identified by organizations. Only a few processes should be validated because of an insufficient ability to measure or test the process outputs and it is due to mostly with the necessity to perform destructive testing [10].

Fig. 5.5 General procedure for validating processes

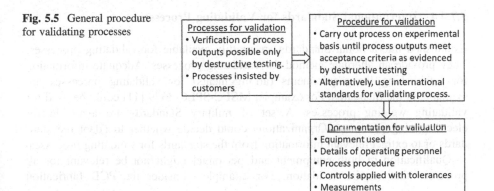

For example, brazing of parts requires validation as the strength of each brazed part can only be determined by destructive testing. In general, a process is identified for validation if the primary characteristic of the output of the process can only be measured through destructive tests.

PCB Assembly process generally does not require validation as inspection and tests could be performed on the assembled PCBs. However, customers might insist for validating (using specific equipment, procedure and qualified persons) the PCB Assembly process for aerospace applications and organization should maintain procedure for validating the process.

5.7.4.3 Procedure for Validation

The requirements (arrangements) that need to be determined and documented for validating processes are:

  (i)   Define criteria for the review and approval of process outputs.
 (ii)   Approval of equipment and qualification of personnel.
 (iii)   Use of specific methods and procedures.
 (iv)   Requirements of records.

Thorough understanding of process is necessary for validating the process. The specifications of product (output of process) are documented before validating process. The specification includes the requirements of both destructive and non-destructive tests. It defines the criteria for the review and approval of process outputs. A process is validated when destructive tests on the resulting product of the process demonstrate compliance to the specifications. Equipment used for the process, operating personnel, process procedure, controls applied with tolerances and measurements are documented. The documented information defines the requirements for validating the process. The requirements of non-destructive tests are finalized after the validation of processes. They are used as verification tests for the outputs of validated process in routine production operations.

5.7.4.4 International Standards for Validating Processes

Established international Standards are readily available for validating processes. They provide an alternate method for validating processes. Adequate information for establishing the requirements (arrangements) for validating processes are available in the Standards. For example, MSFC-SPEC-3679 [11] could be used for validating welding processes. A set of military Standards are available for electro-plating processes. Organizations could decide whether to adapt the standards or to extract relevant information from the standards for validating processes.

Qualification of both equipment and personnel might not be relevant for all processes that require validation. For example, consider the PCB fabrication process. Qualification approval tests are conducted on samples and test coupons as per the Standard, MIL-PRF-31032 [12]. If the samples and test coupons are found satisfactory as per the acceptance criteria of MIL-PRF-31032, the methods and procedures of PCB fabrication process are approved (validated) for production. Operators are trained for carrying out the process and the requirements of records are defined. Approval of equipment and qualification of personnel need not be separately established for PCB fabrication process.

5.7.4.5 Revalidation of Processes

Validation for processes is usually for a period of three years and the period of validity is based on military and other international Standards. Revalidation is performed as per the validation procedure. However, revalidation is repeated for changes either for improvements, changes or obsolescence of equipment, etc.

5.7.5 Preventing Human Errors

Human errors cause failures in the implementation of production processes. Actions to prevent human errors are documented in appropriate generic quality procedures for production processes. Additional information is provided for understanding the sources of human errors.

The potential causes of human errors are improper handling of products, disregarding work rules (instructions), improperly made rework, failures not detected during visual inspection and misunderstanding of process and product requirements [13]. Improper handling of products is unintentional error due to non-availability appropriate handling facilities for products. Disregarding work instructions is intentional error due to lack of enforcement. Improperly made rework might be due to lack of training and non-availability of recommended tools for rework. Failures not detected during visual inspection might be due to non-availability of suitable facilities. Magnifiers and microscopes might have to be replaced by video displays. Training, automation, motivation, providing resources and process audits are some of the actions to prevent human errors. Interlocking human interaction with machines for preventing safety related failures in high voltage applications is also an example of the actions.

5.7.6 Release, Delivery and Post-delivery Activities

The controlled conditions for the implementation of release, delivery and post-delivery activities are documented in appropriate generic quality procedures for production processes. Requirements for post-delivery activities are explained in Sect. 5.11. Requirements for release and delivery activities are explained in Sect. 5.13.

5.8 Identification and Traceability

Two requirements for identification are specified in ISO 9001 and the requirements are product identification and product status identification. Traceability requirement is optional to organizations and if implemented, records should be maintained.

5.8.1 Product Identification

Product manufacturing is a set of many production processes. It begins with planning process and ends with delivering finished products to end customers. Considerable movement and storage of many types of products occur during product realization. Products could be raw materials, semi-finished items or finished products. The products are identified by suitable means during movement and storage. Products are identified when production processes are carried out also on them.

Usually, engineering drawing exists for every type of product. The drawing number of product is used for product identification. Suitable tags indicating drawing numbers or any other suitable method could be used for identifying products during movement and storage.

5.8.2 Product Status Identification

It is not adequate if products are identified during production processes. The product status should also be identified. Product status indicates which production process is completed on the product and whether the product is accepted or nonconforming or rejected or otherwise. Usually, multi-colored tags indicating both product identification and product status are attached with products. Any other convenient method could also be used for identifying product status.

5.8.3 Traceability Requirements

Traceability procedure is implemented by many service organizations such as airlines and postal departments. Industries implement the procedure on their own for identifying the sources of purchased products and internal production processes when field failures are reported by customers. Implementation of the procedure might also be specified in contracts by customers for aerospace and military products. Software system is essential for implementing traceability procedure with appropriate records. Records provide linkages to sources of suppliers, details of production processes and inspection records. It is very common in service organizations and industries to use bar code system for implementing traceability procedure. Bar code system generates unique identification numbers for products. Traceability requirements are not mandatory for organizations.

5.9   Property of Customers and External Providers

Customers might provide special raw materials or components for incorporating the items into their products that are processed by the organization. Tools or test equipment or engineering documents (intellectual property) might also be provided to organization, supporting the processing of the products. It is preferable to maintain documented information for handling customer property. The procedure is applied for handling property belonging to external providers.

The procedures to verify, identify and preserve customer property are same as those explained in Sects. 5.5.5.5, 5.8 and 5.10. The details of contracts for which the customer property is received and customer are included for identifying customer property. Customer property is protected and safeguarded until they are used or incorporated into finished products.

5.9.1   Discrepancies in Property

Verification tests on customer property are visual examination for damages, quantity, checking the receipt of correct item and other tests appropriate to the property. Any discrepancy observed in verification tests or loss or damages when production processes are carried out are recorded and the same are reported to customers or external providers as applicable. Documentary information is retained to provide evidences.

5.10   Preservation of Product

Preservation of product is relevant for raw materials, adhesives, chemicals, electronic components, semi-finished products, finished products and those supplied by customers. The products are stored until production processes begin. There could be waiting period for the products between and within production processes. The characteristics that were verified on the products should be preserved until further manufacturing operations are carried out or the products are delivered to customers. Preservation includes the provisions for handling semi-finished products from one process to next process. A few examples are provided for understanding the needs for the preservation of products:

(i)   Adhesives should be preserved at recommended ambient conditions.
(ii)   Chemicals should be preserved considering safety provisions.
(iii)   CMOS electronic components should be preserved to prevent electro-static damages. The same precautions should be considered while issuing partial quantities.
(iv)   RF Coaxial cables should be preserved in spools to retain impedance characteristics. The same precautions should be considered while issuing partial quantities.
(v)   Rejected products need to be protected from mixing with other products until they are disposed or returned to manufacturers.

The methods of preservation for Products are designed to satisfy the needs of the products. It is preferred to state the methods in the form of displayed procedures for production procedures with appropriate document controls.

5.11   Post-delivery Activities

Providing technical information to customers, obtaining customer feedback and resolving customer complaints are fundamental post-delivery activities for all products. The activities would be adequate for organizations delivering products like electronic components but they are not adequate for equipment delivered to customers.

Field technical assistance for installation, check-ups, warranty work, out-of-warranty work, product disposal and spare parts distribution are the major activities for equipment focusing customer relationship [14]. There could be requirements of training of users at customer premises after the delivery of equipment. Documentary information is maintained for implementing post-delivery activities. The extent of post-delivery is decided considering the type of equipment.

5.11.1   Extent of Post-delivery Activities

Organizations determine post-delivery activities considering the type of products and services. In determining the extent of implementing the post-delivery activities for the products and services, organization considers:

(i) Statutory and regulatory requirements:
   The statutory and regulatory requirements are adhered when implementing post-delivery activities. Directives for managing electronic wastes [15], procedures for the safety certifications of lifts and regulations for wearing personal protective equipment are examples of statutory and regulatory requirements for implementing post-delivery activities.

(ii) Potential undesired consequences:
   Customer dissatisfaction is generally one of the critical undesired consequences and it is considered in planning the extent of post-delivery activities. After-sales influence the relationship to the extent that customers will switch to competitors and furthermore, influence the reputation of the company [14]. The cost of providing post-delivery activities should also be considered as organization might have constraints.

(iii) Nature, use and intended lifetime:
   The nature, use and intended lifetime of products are considered for planning the extent of post-delivery activities. For example, the lifetime of consumer products is less than that of military products and hence the availability of spare parts is planned accordingly.

(iv) Extent of customer requirements:
   For example, customers of consumer products expect prompt services at reasonable cost especially for out-of-warranty period. Hence, wider service network is planned compared to military products.

(v) Customer feedback:
   Customer feedback is used for improving the performance of products and services to customers. Extent of obtaining customer feedback is indicated in the quality procedure for customer satisfaction (Sect. 6.3).

5.12 Control of Changes

Product related engineering documents and generic QMS documents are amended for implementing corrective actions identified during the cause analysis of product and process failures. The changes to the documents are controlled as per the procedure in Sect. 4.8.2. The proposed changes are reviewed for ensuring conformity to products and the appropriate actions for amending documents are implemented by authorized persons. Documented information is retained to provide evidences for the review and the actions arising from the review.

5.13 Release of Products and Services

The inspection and tests are implemented at the appropriate stages of production processes as planned in product related engineering documents. Quality records are maintained for the results of inspection and tests. Products are released for dispatch to customers after the quality records provide evidences for satisfactory results. Organizations usually maintain documented information (quality procedure) for controlling the release of products to customers.

Release of products to customers could be initiated before the completion of planned inspection and tests if the release is approved by customers or their authorized representatives. Documented information (quality records) is retained to provide evidences for the approval by customers.

5.14 Control of Nonconforming Outputs

Nonconforming outputs are those products or services which are found not complying with their relevant engineering documents during production and service provision (Sect. 5.7). Control of nonconforming product is explained for products and it is applicable for nonconforming service also. Nonconforming products could be parts or semi-finished products or finished products.

Documented information is maintained for the control of nonconforming products. Control refers to:

(i) The identification of nonconforming products (Sect. 5.8.2).
(ii) Preventing their unintended use or delivery by having appropriate enclosures or any other suitable arrangement.
(iii) Taking actions based on the nature of nonconformity and its effect on the conformity of products.
(iv) Authority for deciding the actions on the nonconformities.

There might be situations wherein nonconforming products are detected by organization after the delivery of products to customers. The methods of dealing with nonconforming products are explained.

5.14.1 Actions on Nonconforming Products

Nonconforming products might be usable with or without additional efforts. The methods of dealing with nonconforming products detected during the implementation of production processes and after the delivery of products are explained. One or more of the methods could be used for dealing with the nonconforming products and the methods are shown in Fig. 5.6. In addition to using the methods for dealing

Fig. 5.6 Actions on
nonconforming products

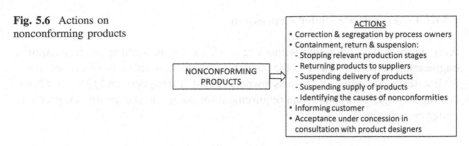

with the nonconforming products, corrective actions appropriate for the nonconformities are also implemented (Sect. 6.8).

5.14.1.1   Correction and Segregation

Nonconforming products could be re-worked to meet the specified requirements
process owners. Conformity to product requirements is verified after re-work.
Nonconforming products could be segregated for defects. After segregation, the
conforming products are accepted. Nonconforming products could be corrected to
meet requirements or scrapped.

5.14.1.2   Containment, Return or Suspension

Containment is identifying the actions required to stop the delivery of known or
suspect noncompliant products and prevent further impact to customer from its use
until the root causes of the noncompliant products are known [16]. Containment
actions are similar to the actions of government departments for treating affected
people under the sudden outbreak of diseases and preventing the spread of the
diseases to others. In industries, the actions could be holding the relevant production stages, suspending the delivery of products to customers and segregating
the stocks in Stores and production stages. Externally provided products could be
returned and suspending the delivery of products is applicable for external providers also. The pace of implementing the actions and identifying the causes of
nonconformities minimizes financial impact on organization. Computerized traceability system supports faster and effective locating of non-compliant products for
segregation.

5.14.1.3   Informing Customer

Customers are informed for re-calling defective products. If appropriate, the stocks
at customer end could be re-called for examination. Informing customers might be a
contractual requirement when non-compliant products are detected.

5.14.1.4   Acceptance Under Concession

Acceptance under concession without re-working nonconforming product requires engineering expertise. Hence, the nature of nonconformities is discussed by a review team consisting of process owners, product designers and QA. Involving customers might be a contractual requirement for accepting non-compliant products under concession.

5.14.2   Documentary Evidences

Documentary information is retained by organization to provide evidences for:

  (i)   Description of nonconformity.
 (ii)   Description of the actions taken.
(iii)   Description of concessions obtained.
(iv)   Identifying the authority for deciding actions in respect of the nonconformity.

References

1. How change is addressed within ISO 9001:2015, International Organization for Standardization, ISO/TC 175/SC2/N1275
2. Natarajan D (2015) Reliable design of electronic equipment: an engineering guide. Springer
3. Velez Cabassa AI (2015) Purchasing controls, FDA small business regulatory education for industry conference
4. Supplier guidebook: guidelines for a successful partnership with Boston Scientific, USA, Aug 2014
5. Supplier quality requirements, United Technologies Aerospace System (UTAS), A9000, Revision J, Sept 2013
6. MIL-STD-810G, environmental engineering considerations and laboratory tests, Jan 2000
7. ASME B31.3-2012, Pressure piping. The American Society of Mechanical Engineers, USA
8. IPC J-STD-001J, Requirements for soldered electrical and electronic assemblies, Feb 2005
9. Schroder PW, Powell DM (2012) CSCMP's a better way to engage with suppliers, Supply Chain, Q4
10. Misztal A (2013) When you need validation of the processes? Mach Technol Mater (11). ISSN 1313-0226
11. MSFC-SPEC-3679, Process specification—welding aerospace hardware, NASA, Oct 2012
12. MIL-PRF-31032B, General specification for printed circuit board/printed wiring board, May 2012
13. Vogt K (2010) Human as an important factor in process production control, industrial engineering. 7th international DAAAM baltic conference, April 2010
14. Egonsson E, Bayarsaikhan K, Ting Ly T (2013) After-sales services and customer relationship marketing: a multiple case study within Swedish heavy equipment machinery industry, Linnaeus University

15. Statutory Instrument No. 149 of 2014, European Union (Waste Electrical and Electronic Equipment) Regulations, 2014
16. Corrective Action Webinar, Topic 5—containment expectations, Lockheed Martin, USA, 2015

Chapter 6
Performance Evaluation and Improvement

Abstract The planning and operational requirements of QMS are implemented for processing customer requirements. Planning is done for processing customer requirements and the outcomes of planning are implemented by the internal operational processes of QMS. The outcomes of the QMS are products delivered to customers, satisfying their requirements. The outcomes of planning and QMS and the performances of the internal processes of the QMS are assessed to evaluate the effectiveness of the QMS of organization. The performance evaluation requirements are monitoring, measurement, analysis, evaluation, customer satisfaction internal audit and management review. The outputs of performance evaluation are the inputs for continually improving the effectiveness of QMS. Adequate information with illustrations is provided for the requirements of performance evaluation and improvement. Key performance indicators are suggested for measuring the performance of QMS processes.

6.1 Introduction

PDCA (Plan-Do-Check-Act) cycle is a continuous improvement tool with four steps. ISO 9001 employs the process approach which incorporates the PDCA cycle and risk based thinking [1]. The processes of ISO 9001 are grouped and shown for each of the four steps of PDCA cycle [2]. The processes of Plan-step of the cycle are explained in Chaps. 3 and 4. The processes of Do-step of the cycle are explained in Chap. 5. The processes of Check-step of the cycle are grouped under performance evaluation and the processes are:

   (i)   Monitoring, measurement, analysis and evaluation
   (ii)   Customer satisfaction
   (iii)   Internal audit
   (iv)   Management review

The processes of Act-step of the cycle are grouped under improvement. The processes under improvement are:

© Springer International Publishing AG 2017   101
D. Natarajan, *ISO 9001 Quality Management Systems*,
Management and Industrial Engineering, DOI 10.1007/978-3-319-54383-3_6

  (i)   General requirements
 (ii)   Nonconformity and corrective action
(iii)   Continual improvement

The processes of performance evaluation and improvement are explained. Examples from industries are provided for understanding the requirements of the processes.

6.2 Monitoring, Measurement, Analysis and Evaluation

Measurement is integral part of monitoring. Analysis and evaluation are part of decision making using measured data. The fundamental objective of monitoring and evaluation of QMS processes is to continually improve the effectiveness of the QMS of organization. General requirements for monitoring and measurement, measuring customer satisfaction, and analyzing and evaluating measurement data are presented. Software support is essential for the implementation of monitoring, measurement and analysis activities associated with QMS processes.

Monitoring and measurement for controlling processes are not presented as they are specific to production processes. Monitoring and measurement for improving the efficiency of processes are also not presented as the processes are identified during management reviews.

6.2.1 General Requirements

Monitoring, measurement, analysis and evaluation activities for QMS processes are planned for implementation. The requirements for planning are:

  (i)   What needs to be monitored and measured
 (ii)   The methods for monitoring, measurement, analysis and evaluation needed
        to ensure valid results
(iii)   When the monitoring and measuring are performed
(iv)   When the resulting data from monitoring and measurement are analyzed and
        evaluated.

Monitoring and measuring the performance of QMS processes generate data. The data is analyzed and the results of analysis are used to evaluate the performance of processes. The evaluation results of processes are used to assess the performance and the effectiveness of the QMS of organization. Appropriate documented information (quality records) is maintained to provide evidences for measurement, analysis and evaluation activities for QMS processes. The planning of monitoring, measurement, analysis and evaluation activities are presented. The general procedure for planning the activities is shown in Fig. 6.1.

Fig. 6.1 Planning of monitoring, measurement, analysis and evaluation

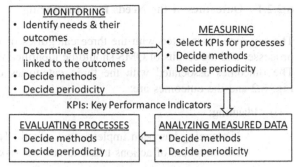

The method of monitoring practically decides the validity of evaluation results

6.2.2 Needs to Be Monitored and Measured

The needs for monitoring are determined from the Plan-Do steps of QMS. The requirements of planning and operation of QMS are implemented as per the QMS of organization for processing customer requirements. The inputs for planning are derived from customer requirements, organization and its context and the needs and expectations of interested parties [1]. Planning is done for processing customer requirements and it includes the resources for operation. The outcomes of planning are the inputs for the internal operational processes of QMS. The inputs are utilized by the internal processes of QMS, transforming the inputs into products. The results (outputs) of the QMS are products conforming to requirements and achieving customer satisfaction. Planning, internal processes and the results of QMS are the needs to be monitored and measured. They are shown in Fig. 6.2.

Fig. 6.2 Needs to be monitored and measured

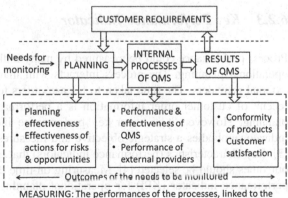

MEASURING: The performances of the processes, linked to the outcomes, are measured using key performance indicators

6.2.2.1 Outcomes of the Needs for Monitoring

The outcomes associated with the three categories of needs i.e. planning, internal processes and the results of QMS should be identified to plan monitoring activities. The outcomes associated with the three categories of the needs are shown in Fig. 6.2 and the outcomes are:

(i) Planning:

   – If planning has been implemented effectively
   – Effectiveness of actions taken to address risks and opportunities

(ii) Internal processes of QMS:

   – Performance and effectiveness of QMS
   – Performance of external providers

(iii) The results of the QMS:

   – Conformity of products and services
   – Customer satisfaction

6.2.2.2 Measuring the Outcomes

One or more QMS processes are linked to the outcomes associated with the three categories of the needs and they are determined to facilitate monitoring. The performances of the QMS processes are monitored and measured using key performance indicators. Key performance indicator is defined before presenting the monitoring and measurement of QMS processes.

6.2.3 Key Performance Indicator

Process performance could be measured by many indicators. Consider the process, operational planning. The process interacts with many other processes including external providers. The time taken to release orders to external providers from the receipt of customer order information is a performance indicator for operational planning. However, it is not a key performance indicator. A key performance indicator embodies a strategic objective and measures performance against a goal [3]. Key characteristics (performance indicators) that have significant impact on the outcomes should be selected [4]. Measuring on-time delivery achievement is a key performance indicator (KPI) for operational planning. The goals for the KPIs of processes and appropriate metrics for the goals are decided by organization. The goals of the KPIs related to statutory and regulatory requirements, are as per applicable Standards. KPIs are decided considering the complexity of products.

Very complex or high-level performance measures may require many raw data from numerous sources [5].

6.2.3.1 Types of KPIs

Two types of KPIs are used for measuring the performance of processes, namely, outcome and driver KPIs; outcome KPIs are used to measure the output of past activity and driver KPIs measure activity in current state [3]. Outcome KPIs are used for evaluating the effectiveness of process outputs. Driver KPIs are suitable for evaluating the effectiveness of planning processes.

6.2.4 Key Performance Indicators for QMS Processes

QMS processes linked to the outcomes associated with the needs of planning, internal processes and the results of QMS of organization are briefly explained. Suggested key performance indicators (KPIs) for the processes are presented.

6.2.4.1 Effectiveness in Implementing Planned Activities

ISO 9001 specifies that organization should plan the activities for delivering results in accordance with customers' requirements and the organization's policies. The planning activities that should be considered by organization are [1]:

(i)   Planning actions to address risks and opportunities
(ii) Design and development planning
(iii) Planning production processes

Monitoring and measurement are performed to evaluate whether the planned activities of processes are implemented effectively. KPIs for measuring the effectiveness of implementing the actions taken to address risks and opportunities are suggested in Sect. 6.2.4.2. KPIs for measuring the effectiveness of implementing design, development and production planning activities are indicated:

(i) Design and development planning:

Monitoring and measuring the completion of sub-systems or their stages of electronic system is one of the KPIs for evaluating the effectiveness of implementing the activities of design and development planning.

(ii) Planning production processes:

Production management and planning is effective when supply matches the demand as closely as possible and the core part of production system is human i.e.

making people work comfortable and happily [6]. The demand and supply between the internal production processes should be met without any gap. Monitoring and measuring the delay and the shortage quantity are the KPIs to evaluate the effectiveness of implementing the activities of operational planning for production processes. Appropriate KPIs could be developed for the human needs in production processes also.

6.2.4.2 Actions Taken to Address Risks and Opportunities

Examples are provided for planning and implementing the actions to address risks and opportunities in design and development (Sect. 3.8) and customer enquiries (Sect. 3.9). Appropriate driver KPIs to measure the time to complete the strategic activities of planning could be developed for evaluating the effectiveness of implementing the activities of the processes. Measuring daily production quantity is another KPI for the actions related to customer enquiries.

6.2.4.3 Performance and Effectiveness of QMS

The internal process approach looks at internal activities and assesses activities by indicators and the approach is important because the efficient use of resources and harmonious internal functioning are ways to measure effectiveness [7]. Performance is measured for the relevant internal operational and support processes of QMS for evaluating the effectiveness of the processes. It underlines the change towards management by information and knowledge instead of primarily relying on experience and judgement [8].

Quality objectives (Sect. 3.12) that are established for the relevant functions, levels and processes of QMS serve as a means to measure the performance of the internal operational and support processes of QMS. The quality objectives are the KPIs of the processes. Appropriate data is generated by monitoring and measuring the key performance indicators for evaluating the performance and the effectiveness of the QMS of organization.

6.2.4.4 Performance of External Providers

Monitoring and measuring the performance of external providers is generally performed for those who contribute 80% of business and those providing critical products. The key performance indicators for the measurement are quality, on-time delivery and service after delivery. Appropriate data is generated by monitoring and measuring the key performance indicators for evaluating the performance of external providers.

6.2.4.5 Conformity of Products and Services

Conformity assessment is any activity which results in determining whether a product corresponds to the requirements contained in a specification [9]. Electrical testing, non-destructive reliability stress screening, computing DPMO (Defects per Million Opportunities) and first pass yield are some of the activities of product quality assurance plan [10]. First pass yield is used as KPI in many industries. Observed defects (discrete data) in products could be expressed as DPMO and the measured values (continuous data) of critical characteristics of products could be expressed as statistical sigma levels. Additional information on DPMO and statistical sigma levels is available in Sect. 5.5.5.5.

6.2.4.6 Customer Satisfaction

Customer satisfaction is one of the results of QMS. Monitoring and measuring customer satisfaction is explained in Sect. 6.3.

6.2.5 Methods for Monitoring and Measuring KPIs

Monitoring, measurement, analysis and evaluation are serially linked activities. The measured performances of processes are analyzed and the results of the analysis are used to evaluate the processes for decision making. The method of monitoring should be defined as it practically decides the validity of results (decisions) when evaluating processes. Guidance is provided for deciding the methods for monitoring and measuring KPIs to ensure valid results.

6.2.5.1 Methods

The method of monitoring is decided by the check points for the monitoring activity. The methods of monitoring and measuring the indicators (KPIs) of processes should be simple, flexible, consistent with the objectives of the processes and able to generate reliable data [4]. The methods of monitoring and measurement are illustrated with examples for better understanding.

Consider an organization with its scope defined as designing, manufacturing and delivering microwave components to customers. The output KPI of operational planning processes is delivery performance. The KPI could be monitored at the dispatch of products to customers. The method of measurement is comparing the dispatch date with the committed delivery date. The monitoring and measurement methods are simple and able to generate reliable data; but they are not in consistent with the objective of evaluating the effectiveness of the operational planning process. Deciding appropriate corrective actions is not feasible if the performance goal of operational planning is not achieved. The method of monitoring should include additional check points such as completion dates for external providers, verification of externally provided products, machining and product assembly. Similarly, the locations selected for monitoring effluent levels decide the effectiveness for evaluating compliance to regulatory requirements.

Additional methods could be used for measuring KPIs. Sensors could be check sheets, micrometer (mechanical measuring gauges), digital voltmeter (electronic test equipment) and automated systems [5].

6.2.6 Methods for Analyzing Data on KPIs and Evaluation

The measurements of performance indicators at the various monitoring points of processes are raw data. They have to be converted into useful information for evaluating the process. The method of analyzing measurement data describes the conversion process. The method could be simple arithmetic expressions or tabulations or application of statistical techniques or specially developed algorithms or graphical representations. Computing DPMO (Defects per Million Opportunities) and statistical sigma levels (Sect. 5.5.5.5) are examples of analyzing data statistically to evaluate the conformity of products. The results of the analysis are quantitatively expressed using appropriate metrics. Reports are prepared for the results of the analysis for processes indicating the goals of the processes.

6.2.6.1 Evaluation Method

The method of evaluating processes defines administrative authorities and the alternatives for decision making. Process owners are responsible for implementing the decisions (results) of evaluating processes. The results of the analysis on performance indicators are used to evaluate the processes. Evaluating a process is assessment followed by decision making. A process is evaluated by:

(i) Assessing whether the process is effective:

This is done by comparing achieved performance level with the goal of the process.

(ii) Deciding appropriate actions for the process:

- Corrective actions.
- Modifying the goals of the process
- New goals for the process

6.2.7 Periodicity of Monitoring and Measuring KPIs

Planning is necessary for how often or when the performance indicators of processes should be monitored and measured. Monitoring and measurement data could be generated on continuous or sampling basis. The frequency of monitoring and measurement is decided by the requirement of evaluating the effectiveness of processes. The frequency with which raw data are collected or measured may have a significant impact upon the interpretation of the performance measure [5]. Higher number of data points might be required for statistical analysis.

Processes are monitored and measured as and when transactions occur for the performance indicators of the processes. For example, the transaction information is dates for operational planning or percent rejections for production processes. Effluent samples are collected (monitored) and the levels as per regulatory requirements are measured once or twice in a day as required. Organization decides the periodicity for monitoring and measuring KPIs.

6.2.8 Periodicity for Analyzing Data on KPIs and Evaluation

Planning is necessary for how often or when the measured data of processes should be analyzed and when the processes should be evaluated using the results of analysis. Periodicity for analyzing measured data on the KPIs of processes and evaluating the processes could be daily, monthly, quarterly, half-yearly or yearly.

Process owners desire to analyze measured data on the KPIs of production processes monthly for initiating appropriate actions and displaying them in shop floors. The measured data on the performance indicators of processes is analyzed half-yearly also and the outputs of evaluating the processes are made available as inputs for management reviews. Measured data on effluent samples are analyzed daily and evaluated for compliance to regulatory requirements. The periodicity for analysis and evaluation could be modified if sufficient data points are not available. Organization decides the periodicity for analyzing data on KPIs and evaluation.

6.3 Customer Satisfaction

Customer is one of the interested parties of organization. Needs and expectations of interested parties are defined in Sect. 3.3 and they are applicable for customers also. Satisfying or exceeding the needs and expectations of customers should be planned for achieving the business strategies of organization. Customer satisfaction is considered to have been achieved when customers perceive that organization has met their needs and expectations. Customer's perception is monitored for

Fig. 6.3 Achieving customer satisfaction

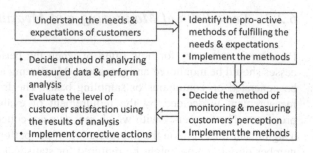

measuring the degree (level) of customer satisfaction. The activities that are planned for monitoring and measurement are shown in Fig. 6.3 and the activities are:

(i)   Understanding the needs and expectations of customers.
(ii) Identifying proactive methods to fulfill the needs and expectations of customers.
(iii) Methods for monitoring and measuring customers' perception
(iv) Methods for analyzing the measurement data
(v)   Evaluating the degree (level) of customer satisfaction using the results of the analysis

6.3.1 Needs and Expectations of Customers

The minimum needs of customers are receiving products as per committed schedule, compliance to product specifications and practically zero field return rate. Customers need prompt actions when field failures are reported. The needs are generally applicable to all categories of products and customers.

6.3.1.1 Expectations of Customers

Fulfilling the expectations of customers significantly contributes for ensuring customer satisfaction. However, certain precautions should be observed by organization in understanding the expectation of customers. Organization should not focus on "what customers want" but rather "what customers value most" to surpass customer expectations [11]. The expectations of customers are listed considering the nature of products and the type of customers. The product could be material, component, hardware or software. The customer could be large organization or individual consumer. Some of the general expectations of customers are:

(i)   Cordial, responsive and accurate communication
(ii) Technical information for product applications
(iii) Extended warranty period

(iv) Reliable performance
 (v) Prompt services
(vi) When relevant for products:

- Aesthetics
- More features
- Availability of spares
- Safety related information

6.3.2 Proactive Methods

A reactive mode of customer service can have a significant negative effect on customer satisfaction and can be exhausting for any organization [12]. Hence most organizations understand the expectations of various categories of customers and proactively integrate the expectations with customer communication process. Implementing the approaches is planned considering the nature of products and the type of the customers.

Customers' data regarding the type of business, names, positions, contact details etc. are collected organized. Customer relationship is established by knowing the data in advance and it supports personalized and effective communication with customers.

Preparing product technical information that would be expected by customers is a proactive approach to ensure customer satisfaction. Technical information might be required by customers for the application of products during design, manufacturing and field use. Relevant documents could be prepared in advance. The documents could be linked to ERP software for instant communication to customers as and when needed.

Companies can significantly influence customer satisfaction by clearly framing expectations and actively delivering what is promised [11]. Organization could specify expectations such as warranty terms, product replacement policies, time to attend or resolve customer complaints, procedures to deal with obsolescence and extent of on-site support. The expectations are communicated in advance to customers and organizations must honor them when needed by customers. The expectations are updated after evaluating the results of customer satisfaction surveys.

6.3.3 Monitoring and Measuring Customers' Perception

Customers' perception is monitored by organizing surveys to obtain feedback from customers. Survey format with appropriate data elements is designed to obtain the expectations on the data elements from customers. The data elements are decided considering:

(i) Present business requirements and future strategies.
(ii) Expectations of customers on service and product application:

Follow-up actions on commitments, resolving technical issues, interactive and clear business communication, response to customer complaints, and functioning of the features of products are examples of the expectations.

(iii) The services that organization desires to focus on improvement.

Customer feedback survey is performed through personal meetings or e-mails or any other convenient means. A cross functional team of organization could also visit customer's premises to obtain feedback. The periodicity of survey could be decided as once or twice in a year.

The data elements of survey format are the performance indicators to measure customer satisfaction. Customer feedback on the data elements is quantified as per pre-determined scale. For example, the scale could be fixed as zero to five. If customer feedback for a data element is excellent, the assigned value is five; if it is satisfactory, the assigned value is four and so on. Customer feedback information might be available during business discussions with customers. They are also documented for evaluation.

6.3.4 Methods for Analysis and Evaluation

Suitable algorithms are developed by organization to analyze the assigned values for the data elements of survey format to compute the achieved level of customer satisfaction. The results could be presented for customers or categories of customers as decided by organization. The results of the analysis are used to evaluate:

(i)   Whether the planned degree of customer satisfaction is achieved
(ii)   Deciding corrective actions, as applicable
(iii)   Implementing the corrective actions through changes in QMS processes and in pro-active approaches.

6.3.5 Practical Considerations

Dispatching products with consistent quality and complying with committed delivery schedule definitely contribute for customer satisfaction. The same will be perceived by customers also and the information is readily available with organization. Hence, the information is suitably weighted and linked to the algorithm for computing overall customer satisfaction.

It is not feasible to monitor and measure the perception of all the customers of organization regarding the fulfilment of their needs and expectations. However, no customer should be served poorly; every customer—regardless of economic worth to the business—has the ability to positively or negatively impact a company's reputation and the key is providing respectable service at a cost that is commensurate with the revenue potential [11]. Organization could decide to monitor and measure the perception of customers, who contribute 80% of their business.

6.4 Analysis and Evaluation

The needs of QMS for monitoring are indicated in Sect. 6.2. Monitoring and measurements are performed for the outcomes associated with the needs by measuring the performances of the QMS processes, linked to the outcomes, using KPIs. The outcomes are:

   (i)   Conformity of products and services:
  (ii)   The degree of customer satisfaction
 (iii)   The performance of effectiveness of the QMS
  (iv)   If planning has been implemented effectively
   (v)   The effectiveness of actions taken to address risks and opportunities
  (vi)   The performance of external providers

Appropriate data and information arising from the monitoring and measurements are analyzed and the results of analysis are used to evaluate the outcomes of QMS processes. The methods of analysis including the application of statistical techniques and evaluation for the outcomes are explained in Sect. 6.2. The evaluation findings are the inputs for management reviews.

6.5 Internal Audit

An audit is a "systematic, independent and documented process for obtaining audit evidence and evaluating it objectively to determine the extent to which the audit criteria are fulfilled" [13]. Examples of audit evidences are quality records and observed discrepancies between work instructions and work performed. Examples of audit criteria are documented procedures and product requirements. Internal audits are conducted by the personnel of organization.

The QMS of organization is set of documents consisting of quality manual, quality procedures, product related work instructions, documents of external origin and quality records for the processes of the QMS. The processes include those specified in ISO 9001 and those determined by the organization needed for the QMS. Internal audits are conducted on the documentation and implementation of QMS and they provide information on:

(i)   Whether the QMS confirms to the requirements of ISO 9001
(ii)   Whether the QMS conforms to the organization's own requirements for its QMS
(iii) Whether the QMS is effectively implemented and maintained

6.5.1   QMS Conformance to ISO 9001 Requirements

Document review is called system audit and it is primarily aimed to determine the level of conformance of the auditee document [14]. The review is also called addressing the adequacy of documentation vis-à-vis ISO 9001 [15]. Internal audit is conducted on the documents of QMS for conformance to ISO 9001.

6.5.1.1   Planning Internal Audit

Internal audit is performed when QMS documents are prepared and when the documents are amended for any reason. Activities are planned for conducting internal audit. If required, a procedure could be established describing the planned activities for performing internal audit on the documents of organization. The planned activities are:

(i)   Audit criteria:

Audit criteria are the requirements of ISO 9001.

(ii)   The scope of the audit:

The scope of internal audit is reviewing the quality manual and procedure documents of organization.

(iii) The method of audit:

The requirements specified in ISO 9001 and those addressed in the documents of organization are examined during the audit.

(iv)   Expected outputs:

The expected outputs of the audit are the requirements that are not addressed and those addressed ambiguously.

(v) Selecting auditors:

The internal audit could be conducted by the nominated representative of top management or by the auditors selected by the representative. The representative should preferably be a certified lead auditor and should have an overview of the QMS processes of organization. The selected auditors are trained by the representative. The management representative ensures that the auditors do not review the documents related to their processes.

6.5.1.2 Performing Internal Audit and Maintaining Quality Records

Internal audit on the documents of organization is performed as per the planned or established procedure. The requirements that are not addressed and those addressed ambiguously are identified and recorded. The results of internal audit are reviewed by the management representative and the QMS documents are amended as needed. Quality records are maintained for performing internal audits, the results of the audits and the amendments.

6.5.2 QMS Conformance to Organization's Own Requirements

Auditing QMS for conformance to organization's own requirements is also a document review. The documents of processes, determined by the organization, and the external documents used by the organization are the basis of internal audit instead of ISO 9001. External documents are used in the design and manufacturing of products to achieve leadership and competitiveness. For example, organization might decide to use IPC-7801:2015, Reflow Oven Process Control Standard, for reflow soldering process. The requirements of the standard are integrated generally with relevant product related work instruction documents by design team. The use of external documents might also be specified by customers or enforced by regulatory bodies. Internal audit is conducted on the relevant documents of QMS for conformance.

6.5.2.1 Planning Internal Audit

Internal audit is performed when the documents are prepared for the processes of organization's own requirements and when the documents are amended for any reason. Activities are planned for conducting internal audit. If required, a procedure could be established describing the planned activities for performing internal audit on the documents of organization. The planned activities are:

(i) Audit criteria:

Audit criteria are the requirements of the documents of processes determined by organization for their own requirements and the external documents used by the organization.

(ii) The scope of the audit:

The scope of internal audit is reviewing the documents of the processes and the external documents.

(iii) The method of audit:

The requirements specified in the external documents and those addressed in the relevant documents of the processes are examined during the audit.

(iv) Expected outputs:

The expected outputs of the audit are the requirements that are not addressed and those addressed ambiguously.

(v) Selecting auditors:

The internal audit is usually assigned to auditors from QA.

6.5.2.2 Performing Internal Audit and Maintaining Quality Records

Internal audit on the documents of the processes is performed as per the planned or established procedure. The requirements that are not addressed and those addressed ambiguously are identified and recorded. The results of internal audit are reviewed by design team and as appropriate by management representative. The documents are amended as needed. Quality records are maintained for performing internal audits, the results of the audits and the amendments.

6.5.3 Audit for Effective Implementation and Maintenance of QMS

The processes of organization are implemented as per their relevant QMS documents. Internal audit is conducted to provide information whether the QMS of organization is effectively implemented and maintained. It is necessary to understand effectiveness in auditing.

6.5.3.1 Understanding Effectiveness in Auditing

Internal audit is performed for processes by verifying compliance to the documented procedures. Observations and examining quality records are some of the methods used for verifying compliance. In addition to auditing QMS processes, interactions between the processes should be audited to provide information on the effectiveness of QMS implementation. The internal supplier-customer relationship (Sect. 1.2) is used to audit for effectiveness in the implementation of QMS. Examining the quality records for a customer order from receipt to dispatch for compliance provides information whether the QMS of organization is effectively implemented. The rejections observed in a process measure the effectiveness of implementing its previous process. However, unbiased analysis should be performed on the rejections to confirm that the rejections are caused by its previous process.

6.5.3.2 Key Requirement of Internal Audit

Auditing for compliance to the QMS is the key requirement of internal audit. Organizing internal audits to verify the effective implementation and maintenance of QMS requires more efforts and additional considerations compared to the audit of QMS documents. An audit program is needed for managing internal audits for verifying compliance.

6.5.4 Audit Program

Audit program is defined as the arrangements for a set of one or more audits planned for a specified time frame and directed towards a purpose [16]. The ultimate purpose or the objective of audit program is to ensure that the QMS of organization is effectively implemented and maintained. Audit program should be planned, established, implemented and maintained for achieving the objective of the program. The requirements of audit program are shown in Fig. 6.4.

Fig. 6.4 Requirements of audit program

Planning is deciding the arrangements i.e. activities and resources for implementing audit program. Audit program is established by preparing one or more quality procedures describing the planned activities, corrections, corrective actions for audit results and quality records. Audit program is implemented as per the established quality procedures. Maintaining audit program refers to initiating appropriate corrections and corrective actions based on the results of implementing the audit program. Implementing corrective actions ensures the effectiveness of QMS implementation. Adequate information is provided for planning and maintaining an audit program. The information could be used for establishing quality procedures for internal audit.

6.5.5 Planning Audit Program

The activities of audit program are planned considering the importance of processes concerned, changes affecting the organization and the results of previous audits as appropriate for the audit program. The standard, ISO 19011 [13] is used as a guide to explain the activities of audit program. The audit program activities that are planned are:

(i)   Responsibilities, frequency and methods for annual audit plan
(ii) Requirements and reporting for each audit:

 - Selection of internal auditors
 - Audit objective, criteria and scope
 - Administrative requirements
 - Reporting audit findings
 - Impartiality and objectivity in conducting audit

(i)   Corrections and corrective actions for audit results
(ii) Quality records

6.5.6 Responsibilities, Frequency and Methods

Internal audits are usually managed by the nominated representative of top management. The representative should preferably be a certified lead auditor, having knowledge on audit principles, procedures and methods. The representative should have an overview of the QMS processes of organization. The responsibilities of the representative are:

(i)   Training auditors within organization
(ii) Preparing annual audit plan:

 - Ensuring that all the requirements of QMS are audited annually.

- Linking the requirements of QMS with the functions of organization in the preparation of annual audit plan.
- Scheduling audits for the functions of organization.

(iii) Planning periodical audit programs as per annual plan:

- Organizing internal auditors
- Defining audit objective, criteria and scope
- Audit administration
- Defining method of reporting audit results, preferably using a pre-designed format.
- Resolving technical ambiguities between auditees and audit teams

(iv) Ensuring the implementations of corrections and corrective actions for audit results.

(v) Maintaining quality records.

6.5.6.1 Frequency of Audit

The results of internal audits are one of the inputs to management reviews, which are conducted twice in a year. The frequency of internal audits is also conducted at least twice in a year. The considerations for deciding the frequency of audit are:

(i) Importance of processes:

- Operational processes of QMS having higher level of functional complexity might require frequent audits.
- Processes requiring compliance to regulatory requirements might have to be audited frequently.

(i) Results of previous audits:

Higher number of nonconformities might have been reported in some of the processes of QMS by internal or external auditors. More number of internal audits is planned for such processes.

6.5.6.2 Audit Methods

Four types of audit methods are commonly used for verifying whether the QMS is effectively implemented. Appropriate methods are selected considering the importance of processes. The methods of conducting internal audit are:

(i)   Interviewing auditees:

Auditees are assessed for their level of understanding procedures through appropriate questions and discussions. Higher level of understanding indicates compliance to the procedures during implementation.

(ii) Observing operations:

It is a direct and practical method to assess the compliance to quality procedures during implementation.

(iii) Examining documents and quality records:

It is a traditional method to assess the compliance to the procedures during implementation.

(iv) Using audit check list:

It is appropriate to use check list for processes that should comply with regulatory and safety requirements.

6.5.7   Planning Requirements and Reporting

Requirements and reporting are planned for conducting each audit as planned in annual audit plan. The planning activities are:

(i)   Selecting internal auditors
(ii) Defining audit objective, criteria and scope
(iii) Administrative requirements
(iv) Reporting audit findings

6.5.7.1   Selecting Internal Auditors

The availability of trained internal auditors is ensured. If required, management representative conducts internal auditor training course for personnel. The training focusses on the objective, criteria and scope of audit and on the internal audit procedure for conducting audits. The availability of technical experts, as needed, for auditing processes associated with complex products is also organized.

6.5.7.2   Audit Objective, Criteria and Scope

Audit objectives are to verify whether processes are implemented as per QMS documents and to record nonconformities, if observed. Audit criteria are the relevant documents of processes to be audited. The documents are quality procedures, engineering documents and external standards as applicable for the processes.

The scope of audit is the processes and functions defined by the representative of management for internal auditors. The scope is defined considering the changes affecting organization. The changes could be addition or omission of processes or changes in the physical locations of functions or in the scope of organization.

6.5.7.3 Administrative Requirements

Administrative arrangements are planned for conducting internal audits. Audits are notified in advance to process owners with schedules. Duration of audits is decided considering the processes and functions allocated to audit teams. Total number of days for audit program is determined. Pre-designed audit formats are provided to internal auditors to control audit process for recording and reporting. The format is designed to record the suggestions auditors for process improvements.

6.5.7.4 Reporting Audit Findings

Internal auditors record the observed nonconformities in the pre-designed audit formats provided by management representative. The nonconformity sheets are returned by internal auditors to the management representative after completing audits. Implementing corrections and corrective actions are organized as per Sect. 6.5.9. Internal audit summary report is prepared using the results of internal audits. The report is one of the inputs for management reviews for evaluating the overall effectiveness of the QMS of organization.

6.5.8 Impartiality and Objectivity in Conducting Audits

Internal auditors should perform audits impartially i.e. free from bias and conflicts of interest. When processes and functions are assigned to internal auditors, the management representative ensures that the auditors do not audit their own processes. Objectivity is ensuring that audit findings and conclusions are based on audit evidences. An example is provided to understand objectivity in conducting internal audits.

6.5.8.1 Objectivity in Internal Audits

Nonconformities, observed during internal audits, are recorded. The recorded nonconformities should be based on objective evidences. The objective evidences are those which could be verified and could direct initiating corrections and corrective actions for the nonconformities. Recording nonconformity as, "The special tool mentioned in the work instruction procedure is not used for tuning and locking RF Filters" does not provide objective evidences. The nonconformity should be recorded as, "Normal screw driver instead of the Torque screw driver, TS456, mentioned in the work instruction procedure, XY123, is used for the assembly of RF Filters, MW528, for the execution of internal work order, WO789". The objective evidences are the references of Torque screw driver, work instruction procedure and internal work order.

6.5.9 Corrections and Corrective Actions

Corrections are the measures implemented to ensure conformity to the specific case of nonconformity observed by internal audit team. Corrections alone might not be adequate for the nonconformities. Additional measures are required to ensure that the nonconformities are not repeated in processes in future. The additional measures are corrective actions.

Recording nonconformities is usually limited to one per sheet. The arrangement facilitates the closure of nonconformity sheets after implementing corrections and corrective actions for the nonconformities. The management representative ensures that corrections and corrective actions are implemented. Management representative distributes the copies of nonconformity sheets to process owners. The sheets indicate the details of nonconformities reported by internal audit team in their processes. Process owners implement appropriate corrections and corrective actions for the nonconformities. The details of the actions are recorded preferably in the same sheets and communicated to management representative. Nonconformity sheets are closed after receiving information on corrections and corrective actions. If appropriate, follow-up audits are conducted to ensure the effectiveness of corrective actions.

6.5.10 Quality Records

Quality records are maintained for competence of management representative, annual audit program, planning actions for performing periodical audits including training records for internal auditors, the results of audits and the implementation of corrections and corrective actions for nonconformities are maintained. The records provide evidences for implementing internal audits.

6.6 Management Review

Top management reviews the QMS of organization at least once in six months. The review demonstrates the commitment of top management for maintaining and improving the effectiveness of QMS. The objective of the review is to ensure suitability, adequacy and alignment with the strategic direction of the organization. The representative of management plans and organizes the inputs for the review in advance. Process owners participate in the review and present the performance status of their processes for review.

6.6.1 Management Review Inputs

Review inputs are prepared by management representative and process owners in summary form, suitable for decision making by top management. The measured performances of support and operational processes including the outputs of QMS are the primary inputs for management review. The summary of internal and external audits is also presented.

6.6.1.1 Actions from Previous Reviews

The outputs of management review are mostly in the form of action points, assigned to relevant process owners with time schedule for completion. The status of the action points of the previous management reviews is discussed until they are closed. Evidences for the implementation of the action points are maintained.

6.6.1.2 Changes in External and Internal Issues

Monitoring the changes in external and internal issues that are relevant to organization is assigned to appropriate process owners during management reviews (Sect. 3.2). The results of monitoring the issues are presented by the process owners during the reviews. The results are discussed and appropriate actions are decided for improving the QMS of organization.

6.6.1.3 Results of Performance Evaluation

Information on the performance and effectiveness of the QMS is discussed in management review. The sources of information are the outputs of monitoring, measurement, analysis and evaluation, customer satisfaction and internal and external audits. The trends in the performance indicators and the outputs of evaluation are presented by process owners for:

(i) Customer satisfaction (Sect. 6.3)
(ii) Feedback from interested parties:

These are usually feedback from customers and external providers. Customer feedback information is obtained from customer satisfaction survey.

Monitoring the needs and requirements of external providers is assigned to appropriate process owners during management reviews (Sect. 3.3). Feedback from external providers is presented by the process owners.

(iii) Quality objectives, process performance, conformity of products and services and performance of external providers with monitoring and measurement results (Sect. 6.2.4).
(iv) Internal audit results (Sect. 6.5) and external audit reports
(v) Performance of external providers (Sect. 6.2.4.4)

6.6.1.4 Other Inputs

The additional requirements for people, infrastructure, operating environment and enhancing organizational knowledge and competence of personnel are presented by process owners for the implementation of the QMS. The effectiveness of the action taken to address risks and opportunities (Sect. 6.2.4.2) are also presented by relevant process owners. The inputs from process owners are reviewed. New opportunities for improving the QMS are identified during the review.

6.6.2 Management Review Outputs

The discussions on the inputs of management review are recorded. Documented information is retained to provide evidences for discussing the management review inputs and for the review outputs. The outputs of the discussions are summarized and presented as decisions and action points, classified as:

(i) Opportunities for improving the effectiveness of QMS
(ii) Needs for changes to the QMS
(iii) Resource needs (people, infrastructure, operating environment and enhancing organizational knowledge and competence of personnel).

6.7 Improvement

Improvement actions represent the Act-step of PDCA cycle. Organization determines and selects available opportunities for improvements. The improvement actions are implemented to meet customer requirements and enhance customer satisfaction. The improvement program includes actions for:

(i) Products and services are improved to meet requirements as well as to address future needs and expectations. Improvement actions are suggested in Sect. 6.7.1

(ii) Undesired effects corrected and prevented or reduced. The methods are explained in Sect. 6.8.

(iii) The performance and effectiveness of the QMS are improved. Suggested methods are presented in Sect. 6.9.

6.7.1 Improvement Actions

Correction, corrective action, continual improvement, breakthrough change, innovation and re-organization are examples of improvement actions. Correction is an action to eliminate a detected nonconformity and corrective action eliminates the cause of a detected nonconformity or other undesirable situation [16]. Corrective actions prevent the recurrence of nonconformities. Continual improvement is a small improvement that could be planned and achieved by everyone in their processes and products at minimal cost. Simplifying process operations and improving tools to eliminate product rejections are examples of continual improvement. Additional information is presented for correction, corrective action and continual improvement in Sects. 6.8 and 6.9.

6.7.1.1 Breakthrough Improvements

Breakthrough improvements are large improvements within a reasonable span of time and they require with management initiative. Planning for achieving six-sigma for the critical characteristics of products by design is an example of breakthrough improvement. Such improvements require group efforts and incur initial cost but they result in the reduction of total cost for organization.

6.7.1.2 Innovations

Innovations are improvements that are achieved by acquiring problem solving skills and systemic thinking, supported by trials and experiments. Innovation requires smaller teams compared to larger groups for breakthrough improvements and it also requires management support. Re-organization is the method used by top management for achieving improvements. Breakthrough improvements, innovation and re-organization are not discussed.

6.8 Nonconformity and Corrective Action

Corrections and corrective actions are necessary when nonconforming products are observed during operational processes. Nonconforming products might be observed during:

(i)   Customer complaints relating to products (Sect. 5.3.1.1)
(ii) The verification of externally provided products (Sect. 5.5.5.5)
(iii) The implementation of production processes (Sect. 5.14)

Implementation of corrections and corrective actions is explained for the nonconforming products reported by customers and those observed during the verification of externally provided items. Corrections for the nonconforming products observed during the implementation of production processes are explained in Sect. 5.14 and hence only corrective actions are explained for them.

6.8.1 Corrections

The nature of actions for correcting nonconformities observed in the products delivered by external providers and those reported by customers is different. Actions on the products delivered by external providers are initiated considering the consequences on subsequent processes. The actions are similar to those explained in Sect. 5.14 and the actions could be:

(i)   Returning the products to the external providers
(ii) Segregation
(iii) Re-working and re-inspection
(iv) Acceptance under concession

Customer complaint could be a false alarm for lack of technical information or real. Hence, the nature of the complaint is examined. If it is a false alarm, appropriate technical information is provided to the customer for closing the complaint. If the complaint is accepted, appropriate actions including replacement are taken resolving the complaints to the satisfaction of customers.

6.8.2 *Corrective Actions*

The method of implementing corrective actions is same for the nonconforming products reported by customers and for those observed during production processes. External providers are responsible for implementing corrective actions for the nonconformities observed during the verification of externally provided products. Organization closes the observed nonconformities after receiving corrective action information from external providers.

The need for identifying corrective action for nonconformity is evaluated as the nonconformity might have been caused by inadvertent errors. If the need for the action is established, corrective actions are identified for nonconformities by:

(i) Reviewing and analyzing the nonconformity
(ii) Determining the causes of the nonconformity
(iii) Identifying multiple corrective actions for eliminating the causes of the nonconformity
(iv) Selecting appropriate corrective actions for implementation
(v) Determining if similar nonconformity exist or could potentially occur in other processes and products for initiating preventive actions for the processes and products

6.8.2.1 Implementing Corrective Actions

The corrective actions for the nonconformity is implemented for processes by amending appropriate product related engineering documents. If necessary, risks and opportunities determined during planning (Sect. 3.8) are updated. The effectiveness of the actions are monitored and reviewed. If the actions are found not effective, the process of identifying corrective actions is iterated.

6.8.3 *Quality Records*

Documented information is retained to provide evidences for

(i) The nature of nonconformities
(ii) The nature of actions (corrections) taken
(iii) The results of corrective actions:

 – Effectiveness of corrective actions
 – Updating actions to address risks and opportunities
 – Amendments to product related engineering documents

6.9 Continual Improvement

Actions are identified to continually improve the suitability, adequacy and effectiveness of the QMS. The actions needs to be planned for implementation thus bridging the Act-step with Plan-step of PDCA cycle. The sources for identifying the actions for continual improvement are:

(i) The results of analysis and evaluation (Sect. 6.4)
(ii) The outputs of management review (Sect. 6.6.2)

Examples of the actions for continual improvements are revising the quality objectives for processes, adding new quality objectives, identifying new KPIs for the processes of QMS, training needs of personnel and improvements in work environment.

References

1. ISO 9001 (2015) Quality Management Systems-Requirements
2. ISO 9001:2008 to ISO 9001:2015, ISO/TC 176/SC 2/N1267
3. Eckerson WW (2009) Performance management strategies, TDWI best practices report, USA, First Quarter
4. Monitoring and Measurement, Module-14, EMS Template, Rev. 2.0, March 2002
5. US Department of Energy (1995) How to measure performance, a handbook of techniques and tools
6. Sato Tomoichi (2008) Production management and planning. JGC Corporation, Interview
7. Bisig E, Szvircsev Tresch T, Seiler S (2007) Improving organizational effectiveness—theoretical framework and model. HFM-163/RTG Working Group, Swiss Military Academy at the ETH, Zurich
8. Phusavat K, Anussornnitisarn P, Helo P, Dwight R (2009) Performance measurement: roles and challenges. Ind Manage Data Syst 109(5):646–664
9. IEC 2008-10, Edition 3.0, Conformity Assessment for Developing Countries
10. Quality Assurance from A-Z, Ericsson, AB2015, March 2015
11. Badgett M, Connar W, McKinley J Customer Satisfaction: Do You Know the Score:?, IBM, G510-1675-00, 06-02
12. Tambellini VT, Liu TSK (2005) Are you exceeding your customers' expectations, strategies, metrics, and best practices for a customer-focused center. An Oracle White Paper
13. ISO 19011:2011, Guidelines for Auditing Management
14. Internal Auditing Course, American Association for Laboratory Accreditation, 2012
15. Cianfrani C, West J (2013) Auditing Process-based quality management systems. ASQ
16. ISO 9000:2005, QMS-Fundamentals and Vocabulary

Chapter 7
Implementing QMS with ERP Software

Abstract The requirements of QMS should be amalgamated i.e. integrated with the operational processes of the organization from the receipt of customer requirements to the delivery of products to customers for achieving and sustaining business growth. Integrating QMS requirements with ERP or other software is the most effective method and it provides valuable benefits to organization. The methods of integrating the QMS requirements with ERP software are illustrated for three operational processes in Chap. 8. Software system engineering is briefly introduced. Requirements analysis with abbreviated activity diagrams and user acceptance testing are presented for providing inputs for developing software for the control of measuring resources.

7.1 Understanding Integrated Approach

Organization establishes ISO 9001 QMS for improving their business prospects and growth. The requirements of the QMS should be amalgamated i.e. integrated with the operational processes of the organization from the receipt of customer requirements to the delivery of products to customers for achieving and sustaining business growth. If the QMS requirements are not integrated with the operational processes, the requirements of the QMS and the schedules of operational processes would compete with each other resulting in conflicts, wastage of resources, customer dissatisfaction and breakdown of the QMS. The integrated approach can be effectively implemented through software.

Organizations use ERP (Enterprise Resource Planning) software or other software systems for processing customer orders from their receipt to completion. Most of the requirements of QMS requirements could be integrated with the ERP software. The integrated approach provides valuable benefits to organization.

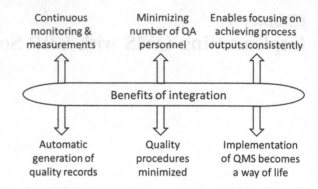

Fig. 7.1 Benefits of integration with ERP software

7.1.1 Benefits of Integrated Approach

The benefits of integrating QMS requirements with ERP software for implementation are shown in Fig. 7.1 and the benefits are:

(i) Process groups could focus on their core tasks for realizing the planned outputs of their processes with consistent quality.
(ii) Monitoring and measurements are performed continuously for evaluating the performances of processes.
(iii) Quality records of processes are generated automatically.
(iv) The number of QA personnel could be minimized.
(v) Preparation of documented procedures could be minimized.
(vi) The QMS of organization is implemented subconsciously by process groups and the implementation of QMS becomes a way of life.

7.1.2 Feasibility of Integration with ERP Software

ERP (Enterprise Resource Planning) is generic system software. However, more and more software companies are starting with a generic system and customizing it to the needs of a particular customer and ERP is an example to this approach [1] The requirements of ISO 9001 QMS are similar to the contractual requirements of organizations for customizing the proposed ERP software. Integrating the requirements of the QMS with the software does not warrant any changes in the design of ERP software. Only additional controls and outputs are needed from the software. Hence, the requirements of ISO 9001 QMS processes except support processes could be added to the set of contractual requirements of organizations for customizing the generic ERP software.

Support processes of the QMS also require software packages. However, it is economical and convenient to develop stand-alone software packages for the support processes, namely, competence of personnel and control of measuring resources. They could be accomplished using Excel with macros also for smaller organizations. The basic concepts of software system engineering are introduced before explaining the integration of QMS requirements with ERP software. The explanation is limited to the extent needed for the integration.

7.2   Software System Engineering

Software system engineering for integrating QMS requirements with ERP software is part of managing ERP project for an organization. The functions of software system engineering are [2]:

(i)   Requirements analysis (requirements of proposed software system)
(ii)   Software design (architectural design)
(iii)   Process planning (strategies and plans)
(iv)   Process control (monitoring activities)
(v)   Verification, validation and testing of system (internal evaluation)

In addition to software system engineering functions, detailed software design, coding and subsystem integration and testing are also required. Finally, users conduct acceptance testing on the software system.

Standard books on software engineering could be referred for understanding software design, process planning, process control, internal evaluation, detailed software design, coding and subsystem integration and testing as the topics are outside the purview of the book. Requirements analysis and user acceptance testing of software system are introduced.

7.2.1   Requirements and Acceptance Testing of Software

Requirements of proposed software system are classified as functional and non-functional requirements [1]. Non-functional requirements are administrative needs regarding hardware, security, etc. and they are not explained. The functional requirements are the technical inputs for developing the proposed software system of organization. The functional requirements are processed by performing requirements analysis and system functional analysis to finalize requirements for developing software system [3].

Acceptance testing of software system is similar to the qualification approval testing of equipment by users. It is performed using an approved acceptance test plan. The test plan is part of contractual requirements for procuring software.

7.3   Integrating QMS Requirements with ERP Software

The quality procedures of QMS describe functional requirements for the operations, monitoring, controls, measurements, outputs and other needs of QMS processes. The "Systems Engineering Best Practices with the Rational Solution for Systems and Software Engineering Deskbook Release 3.1.2" of IBM [3] are applied for processing the functional requirements of QMS processes to integrate the requirements with ERP software. Requirements analysis and limited system functional analysis are performed on the functional requirements of QMS processes.

7.3.1   Requirements Analysis

In requirements analysis, the functional requirements of the quality procedures of QMS processes are transformed into "shall" statements and they are documented as draft system requirements. The requirements analysis is usually performed by organization. The draft system requirements are the inputs for system functional analysis.

7.3.2   System Functional Analysis

In system functional analysis, the draft system requirements are transformed into graphical representation of system operations. Activity diagrams, sequence diagram and state chart diagram are used for the graphical representation [3]. The assistance of software tools is required for performing system functional analysis and the analysis is performed by software project management team. Software tools like **Rhapsody SE-Toolkit** of IBM could be used for system functional analysis for developing software and the graphical outputs are executable for developing software [3]. However, organization should perform limited system functional analysis to transform the draft system requirements of QMS processes into abbreviated activity diagrams.

7.3.2.1   Benefits of Abbreviated Activity Diagram

The graphical representation of draft system requirements of QMS processes using abbreviated activity diagrams is considered part of performing requirements analysis. The diagrams focus on the functional flow of the processes. The benefits of preparing abbreviated activity diagrams are:

(i)   The diagrams function as internal verification tool to eliminate errors in the preparation of draft system requirements.

(ii) The diagrams provide high level of clarity to project management team for capturing the requirements of monitoring and measurements and quality records of QMS processes for integration with ERP software.

(iii) Time to finalize the inputs for the integration is minimized.

7.3.3 Illustrations with QMS Processes

Performing requirements analysis with abbreviated activity diagrams is applicable for developing stand-alone software for support processes and for integrating the requirements of operational processes of QMS with ERP software. Performing the analysis is illustrated for the support process, Control of measuring resources. The analysis is illustrated for three operational processes in Chap. 8.

7.4 Software for the Control of Measuring Resources

Organization maintains one or more quality procedures for controlling its measuring resources. The procedures address the requirements of periodical calibration, maintenance and quality records. The quality records are the master list of measuring resources with relevant calibration and maintenance information. Software support is required for implementing the quality procedure and generating quality records. Requirements analysis with abbreviated activity diagrams and acceptance test plan are presented to develop software for the quality procedures. Editing, password and other protection requirements are excluded to keep the analysis simple.

7.4.1 Requirements Analysis for Operational Needs

The operational requirements of the quality procedure are transformed into "shall" statements in requirements analysis and presented as draft system requirements. Examples are provided for the pertinent data of measuring resources. The draft system requirements are:

(i) Measuring resources database shall be maintained. The pertinent data of each measuring resource shall be equipment data.
Equipment data:

- Description
- ID (serial or control number)
- Make/Model
- Brief capabilities

(ii) Appropriate calibration data shall be updated for each measuring resource by selecting its ID number. The pertinent data of each measuring resource shall be calibration data.
Calibration data:

   – Certificate reference and date
   – Limitations in calibration
   – Calibration due date

(iii) Maintenance history data shall be maintained and updated for each measuring resource by selecting its ID number. The pertinent data of each measuring resource shall be history data.
History data:

   – Date of purchase
   – Calibrated by
   – Nature of adjustments and maintenance, if any.

7.4.1.1   Abbreviated Activity Diagram

The abbreviated activity diagram representing the draft system requirements for the operations of the quality procedure is shown in Fig. 7.2.

7.4.2   Requirements Analysis for Monitoring Needs

The calibration due dates of measuring resources are monitored at the beginning of each month to identify the measuring resources requiring calibration in advance. The draft system requirements after performing requirements analysis on the monitoring requirements of the quality procedure are:

Fig. 7.2 Abbreviated activity diagram: Procedure operation

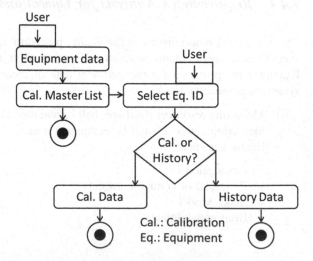

(i) The measuring resources having less than 91 days of valid calibration status as on the date of monitoring shall be identified as calibration due list.
(ii) The measuring resources that are overdue for calibration shall be identified as calibration overdue list.

7.4.2.1 Abbreviated Activity Diagram

The abbreviated activity diagram representing the draft system requirements for the monitoring requirements of the quality procedure to print Cal. Due List and Cal. Overdue List is shown in Fig. 7.3.

7.4.3 Requirements Analysis for Output Needs

Quality records are required for demonstrating compliance to quality procedure. Although electronic records are acceptable, provisions should be available to print records desired by organization. The draft system requirements after performing requirements analysis on the output requirements of the quality procedure are:

(i) Provision to print the calibration master list of measuring resources containing equipment data and calibration data shall be available.
(ii) Provision to print equipment history data of selected ID of the equipment shall be available.

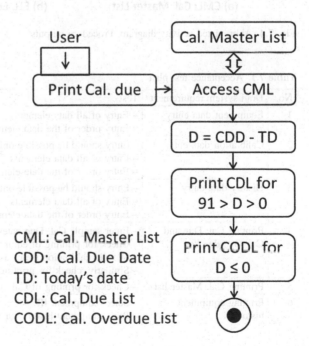

Fig. 7.3 Abbreviated activity diagram: Monitoring outputs

7.4.3.1   Abbreviated Activity Diagram

The abbreviated activity diagram representing the draft system requirements for the output requirements of the quality procedure to print Cal. Master List and Equipment history is shown in Fig. 7.4a, b respectively.

7.4.4   Acceptance Test Plan

Acceptance test plan is prepared to ensure compliance to draft system requirements for the operation, monitoring and outputs of the quality procedure, Control of measuring resources. The test plan is shown in Table 7.1.

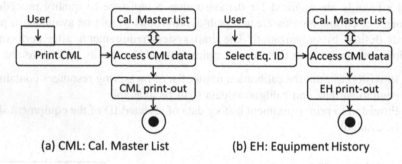

(a) CML: Cal. Master List   (b) EH: Equipment History

Fig. 7.4 Abbreviated activity diagram: Procedure outputs

Table 7.1 Acceptance test plan

No.	Draft system requirement	Test
1	Equipment data entry	– Entry of all data elements – Entry order of the data elements
2	Calibration data entry	– Entry should be possible only after selecting equipment ID – Entry of all data elements – Entry order of the data elements
3	History data entry	– Entry should be possible only after selecting equipment ID – Entry of all data elements – Entry order of the data elements
4	Printing Cal. Due and Cal. Overdue list	– Enter sample Cal. Due dates for selected equipment – Check for printing of equipment having less than 91 days of valid calibration status as on the date of monitoring – Similarly, check for printing of Cal. Overdue list
5	Printing Cal. Master list	– Check for printing of Cal. Master list
6	Printing equipment history	– Select equipment ID – Print equipment history for the selected equipment

References

1. Sommerville I Software Engineering, 8th edn. Pearson Education, UK
2. Thayer RH (2002) Software system engineering: a tutorial. IEEE Computer 0018-9162/02
3. Hoffmann H-P (2011) Model-based systems engineering with rational rhapsody and rational harmony for systems engineering. Deskbook Release 3.1.2. IBM Corporation, USA

Chapter 8
Operational Processes with ERP Software

Abstract Organizations use ERP (Enterprise Resource Planning) software or other software systems for processing customer orders from their receipt to completion. The requirements of operational processes of QMS could be integrated with the ERP software. Integrating QMS requirements is illustrated for three operational processes, namely, Requirements of products, Operational planning and control and Production. Process flow diagrams representing the operations, interactions, monitoring and outputs of processes, requirements analysis with abbreviated activity diagrams and user acceptance test plans are presented for integrating the requirements of the three processes with ERP software.

8.1 Integrating Operational Processes

The functional requirements of support processes (Ex. Control of measuring resources) of QMS could be processed directly by performing requirements analysis and the outputs are documented as draft system requirements. The operational processes of QMS are characterized by interactions with other processes and hence the interactions should be considered before performing requirements analysis.

The requirements of interactions for operational processes might have been documented in different quality procedures. They should be organized for performing requirements analysis. Process flow diagrams are prepared for operational processes to indicate the requirements of the operations, interactions, monitoring and outputs of the processes. The process flow diagrams and the quality procedures of operational processes are used for performing requirements analysis.

8.1.1 Illustrations for Operational Processes

Integrating the requirements of QMS processes with ERP software is illustrated for three operational processes and the processes are:

(i)   Requirements for products and services
(ii) Operational planning and control
(iii) Production and service provision

The illustrations provide the basic approach for integrating the requirements of QMS processes with ERP software. Considerable planning and efforts are required for integrating all the requirements of the operational processes of QMS with the software.

8.1.1.1   Assumptions

In order to keep the presentation of the diagrams and analysis simple, it is assumed that organization designs, manufactures and delivers RF Filters to its customers and the organization has received an enquiry from customer for the requirement of RF Filters. The example is used for presenting the diagrams and analysis to integrate the requirements of the operations, interactions and outputs of the three operational processes with ERP software.

8.2   Requirements for Products and Services

Organization maintains quality procedures for the process, Requirements of products and services (or simply products). The procedures describe the operations, interactions and outputs of the process. The operations are customer communications, determining the requirements of products and reviewing the requirements. The process interacts with planning and design for converting customer requirements into products for delivery. The operations of the process are monitored and measured for evaluating the effectiveness of the process for ensuring customer satisfaction. The outputs of the process are quality records. Process flow diagram, requirements analysis with abbreviated activity diagram and acceptance test plan are presented for the process.

8.2.1   Process Flow Diagram

The process flow diagram for the operations, interactions, and outputs of the process is shown in Fig. 8.1. Monitoring the response time of processing customer enquiries and orders is also shown in the figure.

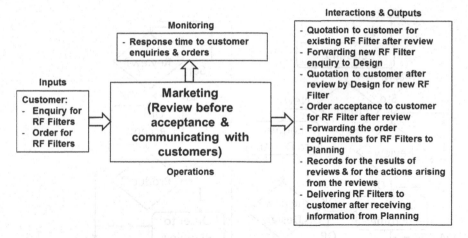

Fig. 8.1 Process flow: Requirements for products

8.2.2   Requirements Analysis

The process flow diagram and the quality procedures are used to perform requirements analysis and for developing abbreviated activity diagram. The draft system requirements for the process are:

   (i) Customer enquiry for the requirement of RF Filters shall be reviewed to decide whether the requirement is standard or new product.

  (ii) If it is standard product, delivery requirement shall be reviewed and if required, it shall be sorted out. If the requirement is acceptable, review record shall be "NIL" and quotation shall be sent to the customer.

 (iii) If it is new product, customer enquiry shall be forwarded to Design for review.

  (iv) If the review by Design team indicates acceptance, quotation shall be sent to the customer as per the information provided by Design; otherwise, non-acceptance shall be indicated. Records for the review by Design shall be maintained.

   (v) When customer order is received, the requirement of RF Filters shall be reviewed to decide whether the order is different from the quotation.

  (vi) Differences, if any, shall be sorted out and records shall be maintained for sorting out. If there are no differences, review record shall be "NIL" and order acceptance shall be sent to the customer.

 (vii) Customer order information for standard and new product shall be sent to Planning. For new product, the order information shall be sent to design also.

(viii) When shipment information for RF Filters is received from Planning, the product shall be delivered to the customer.

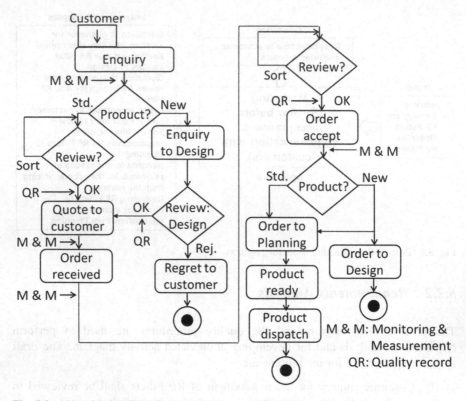

Fig. 8.2 Abbreviated activity diagram: requirements for products

(ix) Response times to customer enquiries and orders shall be monitored and measured and performance evaluation reports shall be generated as per the methods provided by the organization.

8.2.2.1 Abbreviated Activity Diagram

The abbreviated activity diagram for the process is shown in Fig. 8.2. The requirements for monitoring and measurement (M and M) and generating quality records (QR) are shown at the required stages of the activity diagram. It could be observed that the integration of the requirements of the process with ERP software ensures the QMS requirements are implemented subconsciously by marketing team and the quality records and monitoring and measurement for performance evaluation are generated automatically by the software.

Table 8.1 Acceptance test plan: Requirements for products

No.	Draft system requirement	Test
1	Quality records	Check for review records: – After receiving customer enquiry for standard products by Marketing – After receiving customer enquiry for new products by design – After receiving customer order
2	Monitoring and measurement	Response times to customer enquiries and orders and performance evaluation reports as per the methods provided by the organization

8.2.2.2 Acceptance Test Plan

The acceptance test plan is limited to the outputs of monitoring and measurement and quality records. The test plan is shown in Table 8.1.

8.3 Operational Planning and Control

Organization maintains quality procedures for the process, Operation planning and control (or simply planning). The procedures describe the operations, interactions and outputs of the process. The process interacts with marketing, design, external providers and internal production processes for realizing finished products. The operations are planning and monitoring the execution of customer order for RF Filters. The engineering documents of the RF Filter are used for planning. The stocks of in-process, semi-finished and finished products within organization and with external providers are checked for processing the customer order for RF Filters.

Orders are released to internal production processes and approved external providers as needed with appropriate resources and documents. Information to be provided for the orders are documented and controlled by organization. Planning receives externally provided items after verification by QA. Semi-finished and finished products are received from internal production processes after completion. Planning informs marketing when finished products are available for dispatch.

The operations of the process are monitored and measured for evaluating the effectiveness of the process. The outputs of the process are orders to external providers and internal production teams. Orders to external providers and completed internal orders are the quality records. Process flow diagram, requirements analysis and acceptance test plan with abbreviated activity diagram are presented for integrating the requirements of the process with ERP software.

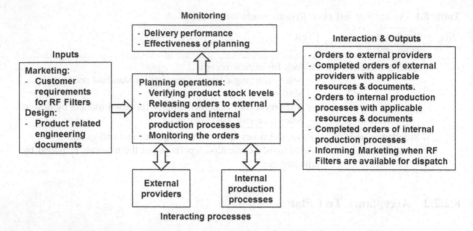

Fig. 8.3 Process flow: Operational planning and control

8.3.1 Process Flow Diagram

The process flow diagram for the operations, interactions, and outputs of the process is shown in Fig. 8.3. Examples of the parameters that could be monitored during the process flow are also shown in the figure.

8.3.1.1 Requirements Analysis

The process flow diagram is used to perform requirements analysis and for developing abbreviated activity diagram. The draft system requirements for operation, interactions, monitoring and outputs are:

  (i) Product related engineering documents shall be used for processing the customer order for RF Filters.
  (ii) The stocks of in-process, semi-finished and finished products within organization and with external providers shall be checked for processing the customer order for RF Filters.
  (iii) The items for further processing shall be identified.
  (iv) Orders to external providers and internal production processes shall be issued with applicable resources and documents.
  (v) Provision for monitoring the orders as per the needs of organization shall be available.
  (vi) Marketing shall be informed when RF Filters are available for dispatch.
  (vii) Orders to external providers and internal production processes shall be preserved.
  (viii) Completed orders from external providers and internal production processes shall be preserved.

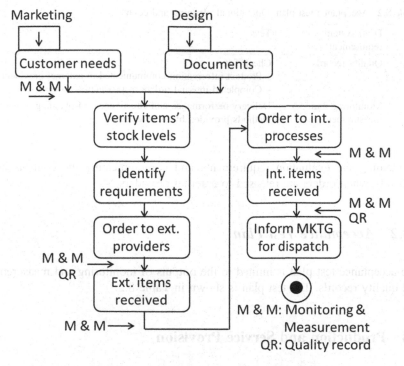

Fig. 8.4 Abbreviated activity diagram: Operational planning and control

(ix) Delivery performance and effectiveness in the implementation of planning shall be monitored and measured and performance evaluation reports shall be generated as per the methods provided by the organization.

8.3.1.2 Abbreviated Activity Diagram

The abbreviated activity diagram for the process is shown in Fig. 8.4.

8.3.1.3 Interaction with QA

Planning interacts with QA for evaluating and re-evaluating external providers. The interaction is not shown in the in the process flow diagram and in the abbreviated activity diagram to keep it simple. External providers are selected after evaluation and the method of evaluation is decided by organization. The type and extent of control on external providers are also decided by organization. The results of evaluating external providers and the controls on them are QMS requirements for the operational planning process. They are linked to product information and to the database of external providers in ERP software. The arrangement ensures

Table 8.2 Acceptance test plan: Operational planning and control

No.	Draft system requirement	Test
1	Quality records	Check for: – Product information communicated to external providers – Completed internal orders of production processes
2	Monitoring and measurement	Delivery performance and effectiveness of planning as per the methods provided by the organization

maintaining the evaluation requirements and communicating the controls automatically when orders are released to external providers.

8.3.2 Acceptance Test Plan

The acceptance test plan is limited to the outputs of monitoring and measurement and quality records. The test plan is shown in Table 8.2.

8.4 Production and Service Provision

Organization maintains quality procedures for the process, Production and service provision (or simply production). The procedures describe the operations, interactions and outputs of the process. The process interacts with planning and control of nonconforming outputs. The inputs for production processes are orders with applicable resources and documents and they are provided by planning. The operations of production processes are performed as per the controlled conditions defined in the documents. The outputs of the processes are semi-finished or finished products and completed orders. They are returned to planning for further processing.

The output performance (quality objectives) of production processes are monitored and measured for evaluating the effectiveness of the process. Nonconformity reports are generated if the planned performance levels of the processes are not achieved. Process flow diagram, requirements analysis with abbreviated activity diagram and acceptance test plan are presented for integrating the requirements of the process with ERP software.

8.4.1 Process Flow Diagram

The process flow diagram for the operations, interactions, and outputs of the process is shown in Fig. 8.5. Monitoring the quality objectives of processes is also shown in the figure.

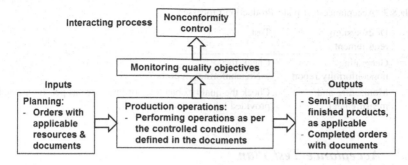

Fig. 8.5 Process flow: Production

8.4.2 Requirements Analysis

The process flow diagram is used to perform requirements analysis and for developing abbreviated activity diagram. The draft system requirements for operation, interactions, monitoring and outputs are:

(i) The achieved outputs of production processes shall be recorded after the completion of production processes.
(ii) The output performance (quality objectives) of production processes shall be monitored and measured and performance evaluation reports shall be generated as per the methods provided by the organization.
(iii) Nonconformity reports shall be generated if the planned performance levels of the processes are not achieved.

8.4.2.1 Abbreviated Activity Diagram

The abbreviated activity diagram for the process is shown in Fig. 8.6.

Fig. 8.6 Abbreviated activity diagram: production

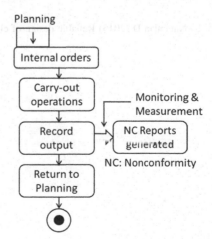

Table 8.3 Acceptance test plan: Production processes

No.	Draft system requirement	Test
1	Generating nonconformity report	Check for: – Nonconformity reports
2	Monitoring and measurement	Check the quality objectives of processes as per the methods provided by the organization

8.4.3 Acceptance Test Plan

The acceptance test plan is limited to the outputs of monitoring and measurement and quality records. The test plan is shown in Table 8.3.

8.5 Design and Development of Products

ERP software is procedure based system. Knowledge based system is required for the design and development of products. For example, the system is for the design and development of electronic products contains product design approaches, the collection of pertinent information from the catalogs of component manufacturers, the application notes of devices, military standards and relevant literature supporting the design of products. The approaches for establishing software system for the design and development of electronic products are explained in the reference [1]. The design and development outputs are finalized engineering documents for products and the documents are linked to ERP software.

Reference

1. Natarajan D (2015) Reliable design of electronic equipment: an engineering guide. Springer

Chapter 9
QMS Planning with Indian Classic, Thirukkural

Abstract Thirukkural is an Indian classic literature in Tamil, one of the ancient languages of India. The classic literature is generally accepted as more than 2000 years old. The literature is secular and some of the earliest translations in French (1848 & 1889), German (1856) and English (1886) indicate its universal acceptance. The ideas of Thirukkural are applied for planning the actions to address risks and opportunities in QMS processes, design and development of products and production processes.

9.1 Indian Classic, Thirukkural

Thirukkural is an Indian classic literature in Tamil, one of the ancient languages of India. It is written by Saint Poet Thiruvalluvar who lived with his wife, Vasuki, in the Indian state of Tamil Nadu. The period of Thiruvalluvar is a subject of discussion. Encyclopedia Britannica mentions the period of Thiruvalluvar as "flourished *c.* 1st century BC or 6th century AD, India". Rev. Dr. G.U. Pope mentions in his book [1] that Thiruvalluvar probably flourished between A.D. 800 and 1000. The classic literature is generally accepted as more than 2000 years old. The verses of Thirukkural were written on palm leaves.

9.1.1 Contents of Thirukkural

Thirukkural contains 133 chapters and the chapters are titled. Each chapter contains 10 verses. The verses of Thirukkural contain just two lines and hence the verses are also called couplets. Kural is the Tamil name for the verses of Thirukkural.

Thirukkural is divided into three parts, as Part-I, Part-II and Part-III. The parts are:

(i) Part-I contains chapters 1 to 38 and the chapters are on Virtue.
(ii) Part-II contains chapters 39 to 108 and the chapters are on Wealth.
(iii) Part-III contains chapters 109 to 133 and the chapters are on Love.

© Springer International Publishing AG 2017

D. Natarajan, *ISO 9001 Quality Management Systems*,
Management and Industrial Engineering, DOI 10.1007/978-3-319-54383-3_9

The three parts of Thirukkural on virtue, wealth and love guide towards happy and prosperous living. Dedicated books on Thirukkural could be referred for more detailed explanation for the three parts. Thirukkural is secular and the translations in various languages indicate its universal acceptance.

9.1.2 Translations of Thirukkural

Apart from translations in many Indian languages, Thirukkural is translated in many languages of the world [1]. Some of the earliest translations in French, German and English are:

  (i) Ariel, M, Kural de Thiruvalluvar (traduits du tamoul), Paris, 1848.

  (ii) Karl Graul, Der Kural Des Tiruvalluver: Ein Gnomisches Gedicht Uber Die Drei Strebeziele Des Menschen (German Edition), Kessinger Publishing, LLC, 1856. Dr.Graul, a devoted student of Tamil literature, published an edition in Leipzig and in London, in 1856 with German and Latin translations [2].

  (iii) Rev. Dr.G.U.Pope, The 'Sacred' Kurraḷ of Tiruvalluva-Nayanar, with Introduction, Grammer, Translation & Notes, Published by W.H.Allen & Co., London, 1886. It may be noted that Kural is spelled as Kurraḷ with double r and the l with a dot below in the title of the book. The spelling conveys that the letters r and l should be stressed when pronouncing the word, Kural.

  (iv) G de Barrigue de Fontainieu, Le Livre de l'Amour De Tirouvallouva, Published by Alphonse. Lemerre, Paris, 1889.

9.1.3 Thirukkural for ISO 9001 QMS Planning

Although Kurals are couplets, every word is packed with useful information. A Kural is a couplet containing a complete and striking idea expressed in a refined and intricate metre [1]. The ideas of the Kurals in Part-II on Wealth could be applied for planning, achieving managerial effectiveness and organizational staffing in industries. Considering the scope of this book, only the applications of the Kurals for ISO 9001 QMS planning are presented.

9.1.3.1 Translation and Commentary

The English translations of Kurals with the title of chapters by Rev. Dr. G.U. Pope [1] and the commentaries for the Kurals from the voluntary body [3] are used for explaining the applications of Kurals for QMS planning. The translation and the

Part I: Virtue	Chapter-1: The Praise of God
Translation of Kural #001	
A, as its first of letters, every speech maintains	
The "Primal Deity" is first through all the world's domains	
Commentary	
As the letter A is the first of all letters, so the eternal God is first in the world	

commentary are shown for the first Kural of chapter-1 for demonstrating the universal acceptance and secularism of Thirukkural. The translation and the commentary of the Kural are self-explanatory.

9.2 QMS Planning Requirements

ISO 9001 specifies that organization should plan the activities for delivering products in accordance with customers' requirements and the organization's policies. The activities that should be considered by organization are [4]:

(i)   Planning actions to address risks and opportunities
(ii) Design and development planning
(iii) Planning production processes.

Planning is deciding activities and resources before implementing the activities to achieve the desired goals of organization. The Indian classic, Thirukkural, specifies the considerations for decision making during the planning processes. The Kurals that are relevant for the planning processes are explained. Some of the Kurals are appropriate for planning actions to address risks and opportunities. Some of the Kurals are appropriate for planning design and development and production processes.

9.3 Planning Actions to Address Risks and Opportunities

ISO 9001 specifies that organization should plan actions to address the risks and opportunities of QMS processes. Three examples are provided in chapter-3 for identifying and integrating the actions with QMS processes. The example provided in Sect. 3.9, risks and opportunities in customer enquiries, is briefly reproduced for understanding the application of the Kurals.

Consider an organization with its scope defined as designing, manufacturing and delivering microwave components to customers. It is assumed that the organization has received an enquiry from customer for delivering large quantity of standard microwave component with relatively with a short delivery schedule, not compatible to the quantity. Although risks are associated in accepting the delivery schedule, the enquiry is treated as opportunity, considering the expectations of interested parties (employees and investors). The actions to address the opportunity are planned to manage the risk before sending the quotation for the customer enquiry.

9.3.1 Thirukkural for Planning the Actions

The information provided in three Kurals are relevant for planning the actions to address the risks and opportunities of QMS processes. The three Kurals are presented before explaining them.

Part II: Wealth	Chapter-48: The Knowledge of Power
Translation of Kural #471	
The force the strife demands, the force he owns, the force of foes	
The force of friends; these should he weigh ere to the war he goes	
Commentary	
Let (one) weigh well the strength of the deed (the purpose to do), his own strength, the strength of his enemy, and the strength of the allies (of both), and then let him act	

Part II: Wealth	Chapter-48: The Knowledge of Power
Translation of Kural #473	
Ill-deeming of their proper powers, have many monarchs striven	
And midmost of unequal conflict fallen asunder riven	
Commentary	
There are many who, ignorant of their (want of) power (to meet it), have haughtily set out to war, and broken down in the midst of it	

Part II: Wealth	Chapter-68: The Method of Acting
Translation of Kural #676	
Accomplishment, the hindrances, large profits won	
By effort: these compare,- then let the work be done	
Commentary	
An act is to be performed after considering the exertion required, the obstacles to be encountered, and the great profit to be gained (on its completion)	

Management personnel and the owners of QMS processes are equated to the kings and monarchs, mentioned in the Kurals. They are responsible for planning the actions to address the risks and opportunities for delivering the microwave components as per customer requirements. The inputs for deciding the actions are explained using the Kurals.

9.3.1.1 Strength of the Deed (Kural #471)

The organization should consider the strength of the deed i.e. accepting to deliver large quantity of microwave component with delivery schedule, not compatible to the quantity. The gap between the strength of deed and the level of efforts of organization to accomplish it represents the risk in executing the deed. Although risk exists in accepting to deliver large quantity of microwave component at the prima facie, the gap between the strength of the deed and the level of efforts is not insurmountable. However, if the organization involved in the design and manufacture of microwave components decides to quote for delivering electronic systems, it might lead to the failure as the strength of the deed is considerable compared to the level of efforts of the organization.

9.3.1.2 Strength of His Own and Allies (Kural #471)

Strength of one's own and allies represents the power of the QMS process owners and their team members in the context of organization. The power of personnel in organization is represented as [5]:

$$Power = Legitimate\ (authority) + Expert + Referent\ (popularity)$$

Out of the three components of power, the expert power of owners and their team members is relevant for planning the actions to address the risks and opportunities. Expert power is referred as the competence of personnel in ISO 9001 and it is based on the basis of education, training or experience appropriate for executing the planned actions. The competence of QMS process owners and their team members is considered for deciding the actions to address the risks and opportunities.

9.3.1.3 Strength of Enemy (Kural #471)

Strength of enemy is equated to the weaknesses of organization. In addition to considering strength, the weaknesses of organization should also be considered in for planning the actions to address the risks and opportunities for QMS processes. Actions such as training, increasing the number of team members and recruiting personnel with appropriate expertise are planned when addressing the actions.

9.3.1.4 Ignorant of Power (Kural #473)

The owners of QMS processes might decide the actions to address the risks and opportunities for the processes using their mental judgement ignoring the need to assess the expertise (power) of theirs and their team members. Such decisions result in failures half way through the implementation of the actions. It is obvious that the failures will cause losses to organizations.

9.3.1.5 Exertion and Obstacles (Kural #676)

The actions to address the risks and opportunities of QMS processes are decided after considering:

(i)   The efforts required for implementing the actions
(ii) The obstacles that are likely to be encountered during the implementation of the actions
(iii) The benefits expected for organization after implementing the actions.

9.4 Design and Development Planning

Organization initiates the design and development planning of new products as per customer orders and at the directives of top management. Some of the activities during the planning are finalizing design and development inputs, deciding the approaches for product design and evaluation of the proto-models of the products. Project manager finalizes the design and development plan for subsequent implementation. The decision making process to finalize the plan is important as it decides the reliability, produceability and total cost of the products. Thirukkural provides practical and useful inputs for the decision making process.

9.4.1 Thirukkural for Design and Development Planning

Two Kurals provide practical managerial information for decision making during design and development planning. The two Kurals are presented before explaining them.

Part II: Wealth	Chapter-47: Acting after due Consideration
Translation of Kural #461	
Expenditure, return, and profit of the deed	
In time to come; weigh these- than to the act proceed	
Commentary	
Let a man reflect on what will be lost, what will be acquired and (from these) what will be his ultimate gain, and (then, let him) act	

Part II: Wealth	Chapter-64: The Office of Minister of state
Translation of Kural #631	
A minister is he who grasps, with wisdom large	
Means, time, work mode, and functions rare he must discharge	
Commentary	
The minister is one who can make an excellent choice of means, time, manner of execution, and the difficult undertaking (itself)	

9.4.1.1 Considering Ultimate Gain (Kural #461)

Ultimate gain is related to total cost in the context of organization. The total cost of ownership model structures with acquisition cost and various operating expenses for products [6]. From the model, the critical components of total cost of product are the cost of design and the cost of correcting failures during manufacturing and usage. Total cost is primarily decided by design and development planning. Direct and support costs are associated with the design and development of products. Decisions for direct and support costs should be made considering the ultimate gain for organization. Two examples are provided illustrating the decisions for the costs.

Direct costs are primarily the cost of design, procuring parts and developing proto-models of the products. Various approaches are identified for the design of products. Some of the approaches might require additional cost (direct cost) for design but the design might result in reliable products, saving the cost of correcting field failures. The approach that results in ultimate gain to organization is decided considering the cost of design, reliability and other requirements of products.

Product design requires support activities such as access to reputed libraries across the world, membership in professional bodies and attending specialized training programs. Although the support costs are not directly utilized for product design, they are essential for developing current and competitive products. Hence, support costs are planned considering ultimate gain instead of considering direct cost of product in isolation.

9.4.1.2 Attributes for Design and Development Planning (Kural #631)

The design and development of new product is usually entrusted to a team of engineers. One of the team members is designated as project manager, who assigns the design and development of sub-systems of the product to the engineers of the team. Project manager requires a plan for monitoring and controlling the design and development of product. The design and development plan is not imposed by project manager but the plan is finalized using the plans of the subsystems, prepared by the engineers. The Kurals provide managerial guidance for the engineers to prepare the design and development plans for their sub-systems and the engineers are equated to the ministers of project manager.

Engineers consider the attributes, mentioned below, for preparing the design and development plans for their sub-systems:

(i) Difficult undertaking:
Difficult undertaking represents the efforts required for the design and development of sub-systems. This is evaluated and understood for preparing realistic design and development plan for the sub-systems.

(ii)   Choice of means:
       Various design approaches exist to achieve the specified requirements of
       sub-systems. They are examined and the relevant approach is selected con-
       sidering performance, reliability and other requirements of project.
(iii) Manner of execution:
       The methods to implement the finalized approach for the sub-systems are
       decided. The methods are used to identify resources and to estimate time for
       the design and development of sub-systems.
(iv)   Time:
       The time that would be required for the design and development of
       sub-systems are estimated and documented.

9.5   Production Planning

Production planning is essential for converting customer orders into deliverable
products. Production planning is a set of managerial and technical activities. One of
the activities of production planning is to ensure the availability of product related
work instructions for carrying out production processes. The applications of
Thirukkural for preparing the work instructions are explained.

9.5.1   Thirukkural for Preparing Work Instructions

A series of production processes are required for the manufacture of products and
product related work instructions are made available for the production processes.
The work instructions specify the controlled conditions for the production processes
to ensure consistent outputs from the processes. Resources, methods, monitoring,
measurements, etc. are the controlled conditions for carrying out the processes. Two
Kurals provide practical considerations for preparing product related work
instructions required for production planning. The two Kurals are presented before
explaining them.

Part II: Wealth	Chapter-47: Acting after due Consideration
Translation of Kural #468	
On no right system if man toil and strive	
Though many men assist, no work can thrive	
Commentary	
The work, which is not done by suitable methods, will fail though many stand to uphold it	

Part II: Wealth	Chapter-68: The Method of Acting

Translation of Kural #675
Treasure and instrument and time and deed and place of act
These five, till every doubt remove, think o'er with care exact
Commentary
Do an act after a due consideration of the (following) five, viz. money, means, time,
execution and place

9.5.1.1 Inappropriate Methods (Kural #468)

The owners of production processes and their assistants are responsible to implement the methods specified in the product related work instruction documents for achieving the specified outputs from the processes. Inappropriate methods might have been specified in the work instruction document for carrying out the production process and such technical errors do occur in industries for various reasons. The production team faithfully implements the work instructions. However, the specified process outputs cannot be achieved by implementing the specified inappropriate methods for the process, even if all the members of the production team strive for the process outputs. In such situations, corrective actions are implemented by amending the work instruction documents to specify the right methods for the production process.

9.5.1.2 The Five Considerations (Kural #675)

The five considerations mentioned in the Kural are applicable before doing any category of act. However, they are more appropriate for preparing product related work instructions as they define the controlled conditions for the work instructions. The work instructions for production processes are prepared considering the following:

(i) Money:
   The primary costs of carrying out production process are the cost of performing operation and inspection. Money refers to the cost of carrying out the process and it is necessary to minimize costs when preparing product related work instructions for production processes.

(ii) Means:
   Means i.e. resources are required both for performing operation and inspection of production processes. Machines, tools, test equipment, etc. are defined in the product related work instructions for production processes.

(iii)   Time:
        Act should be performed at appropriate time. Monitoring, measurement, inspection and testing activities during production process are specified at appropriate stages (time) of production processes in product related work instruction documents for realizing planned process outputs with consistent quality.
(iv)   Methods (for execution):
        The methods (procedures) of executing i.e. carrying out production processes are documented both for operation and inspection.
(v)   Place:
        Like time, act should also be performed at appropriate place. Place refers to the suitable operating environment for production processes and it is defined in work instruction document for carrying out operations and inspections for relevant processes. Suitable operating environment could be controlling temperature, dust level, etc. as required for production process.

References

1. Rev. Dr. Pope GU (1886) The 'Sacred' Kurraḷ of Tiruvalluva-Nayanar, with Introduction, Grammer, Translation & Notes. Published by W.H. Allen & Co., London
2. Translation of the Tamil literary work Thirukkural in world languages & the electronic text of the work on the Net, Webpages of Tamil Electronic Library © K. Kalyanasundaram, http://tamilelibrary.org/teli/thkrl.html
3. http://www.projectmadurai.org/pm_etexts/pdf/pm0153.pdf, Project Madurai (Voluntary body), Thirukkural, English Translation and Commentary by Rev. Dr. G U Pope, Rev. W H Drew, Rev. John Lazarus and F W Ellis
4. ISO 9001:2015, Quality management systems-Requirements
5. Baron R (1983) Behavior in organization: understanding and managing the human side of work. Allyn and Bacon, Inc., USA
6. Lycette B, Lowenstein D (2010) The real "total cost of ownership" of your test equipment. Agilent Technologies, USA, 5989–6642EN

Index

© Springer International Publishing AG 2017
D. Natarajan, *ISO 9001 Quality Management Systems,*
Management and Industrial Engineering, DOI 10.1007/978-3-319-54383-3

Printed in the United States
By Bookmasters